Fisheries Science: A Global Assessment

Fisheries Science: A Global Assessment

Editor: Rory Curtis

R CALLISTO REFERENCE

www.callistoreference.com

Callisto Reference,
118-35 Queens Blvd., Suite 400,
Forest Hills, NY 11375, USA

Visit us on the World Wide Web at:
www.callistoreference.com

ISBN: 978-1-64116-107-7 (Hardback)

Cataloging-in-Publication Data

Fisheries science : a global assessment / edited by Rory Curtis.
   p. cm.
Includes bibliographical references and index.
ISBN 978-1-64116-107-7
1. Fishery sciences. 2. Fisheries. 3. Fishes. I. Curtis, Rory.
SH331 .F57 2019
639.2--dc23

Table of Contents

Permissions

List of Contributors

Index

Preface

Fisheries science is the scientific study of fisheries including their management and development. Fisheries science is a multidisciplinary field that integrates the concepts and principles from a range of subjects such as limnology, oceanography, freshwater biology, marine biology, etc. Fisheries research involves collection of water samples from different depths, acoustic fish finding techniques, etc. to develop a better understanding of the ecological aspects of fisheries. Fisheries can be categorized as saltwater, wild, freshwater or farmed. Tuna, cod, shrimp, crab, oyster, etc. are some of the commonly harvested species. This book essentially covers the crucial concepts related to this field while also discussing the recent advancements made in this field. It is compiled in such a manner that it will provide in-depth knowledge about the theory and practice of fisheries science. This book is an essential guide for both academicians and those who wish to pursue this discipline further.

This book unites the global concepts and researches in an organized manner for a comprehensive understanding of the subject. It is a ripe text for all researchers, students, scientists or anyone else who is interested in acquiring a better knowledge of this dynamic field.

I extend my sincere thanks to the contributors for such eloquent research chapters. Finally, I thank my family for being a source of support and help.

Editor

INDUCED BREEDING OF ENDANGERED STRIPED DWARF CATFISH (*Mystus vittatus*) AND ITS EMBRYONIC AND LARVAL DEVELOPMENT

Rabeya Yesmin[1], Salina Akhter Sume[1], Md. Nazmul Haque[1], Nargis Sultana[2] and Golam Quader Khan[1*]

[1]Department of Fisheries Biology and Genetics, Bangladesh Agricultural University, Mymensingh-2202, Bangladesh, [2]Department of Fisheries Biology and Aquatic Environment, Bangabandhu Sheikh Mujibur Rahman Agricultural University, Salna, Gazipur-1706, Bangladesh

*Corresponding author: Golam Quader Khan· e-mail: khanmgq@yahoo.com

---

ARTICLE INFO	ABSTRACT

Key words:

Mystus vitttus
Induced breeding
Larvae fertilization

The present study reports successful induced breeding of endangered striped dwarf catfish *Mystus vittatus* and its different embryonic and larval developmental stages. Three different doses of PG were tested, viz. 17, 15 and 13 mg PG/kg body weight for female and 14, 12 and 10 mg PG/kg body weight for male with maintaining (1:1) male and female ratio. The hormone doses 13 mg/kg for female and 10 mg/kg for male provided the best result i.e. $91.33\pm2.08\%$ fertilization and $85.00\pm2\%$ hatching rates. Mean survival percentage of the spawns up to 21 days was $8.00\pm1\%$. The fertilized eggs were found to be transparent, demersal, spherical, adhesive and brownish in colour and first cleavage took place within 35-40 min post-fertilization at 29.56 ± 0.25oC. Hatching took place at 24 h. after fertilization. Newly hatched larvae were 3-4 mm in length and slender, transparent and the yolk sac oval in shape. Anus was situated at almost mid ventrally. Larvae started to feed at 48-72 h post-hatching.

INTRODUCTION

Mystus vittatus (Bloch, 1794) is an indigenous catfish species of Bangladesh that belongs to the family Bagridae of the order Siluriformes. It has been listed as an endangered fish species along with the 54 fish species of the inland waters of the country due to overexploitation, aquatic pollution, spread of disease, uncontrolled introduction of exotic fishes, and habitat modification due to industrialization, river-valley projects, excessive water abstraction and siltation due to clearing, the development of breeding technique and so on (IUCN, 2000). The species is locally known as "Tengra" and regarded as a freshwater Small Indigenous Species (SIS) that commonly occurs in inland water areas throughout Bangladesh (Perennou and Santharam, 1990). This SIS contain many minor and trace elements including sodium, potassium, calcium, iron, iodine, zinc, magnesium and phosphorus (Roos et al., 2003). Roos et al. (2003) reported that floodplain fisheries were the main source of fish eaten by rural people of Bangladesh, with SIS contributing the most. They also reported that SIS also dominated the total fish intake in terms of amount and as well as frequency indicating great economic important. Despite its great potential, *M. vittatus* did not receive sufficient attention in aquaculture. Considering its increasing demand and great potentiality, there is a need to start its seed production.

To ensure the availability of fry this species for aquaculture as well as to prevent a fish species from extinction, it is vital to establish a dependable induced breeding and larvae rearing technique. Like other indigenous fish species Tengra is in current threat of extinction for this why to prevent them from extinction and promotion the culture of this species on commercial basis   the technique of induced breeding and larvae rearing is badly needed for our commercial aquaculture on an urgent basis and considering this value the present experiment has been carried out.

MATERIALS AND METHODS

Sample collection and rearing

Fish sample were collected from different locations (Figure 1 and Table 1) in Bangladesh during March to August, 2012. Broodstock domestication was done into the Fisheries Faculty Field Laboratory Complex, Bangladesh Agricultural University, Mymensingh-2202, Bangladesh. Live sample was stocked and reared in the previously prepared separate three rectangular ponds (size 18×14 m^2 and average depth of 1.3 m) of Field Laboratory Complex, Faculty of Fisheries with acclimatization. A special feed (vitamin premix) was applied to fish at the rate of 7-8% of their body weight twice a day (Table 2). In every week growth was measured in terms of length and weight (g). Brood fish are rearing up to their sexual maturation.

Breeding trail

After brood fish rearing up to their sexual maturation and breeding was conducted in fiber plastic tank (1 x 2 x 1 ft^3) using ready to breed fish and physico-chemical conditions of water as follows temperature, DO and pH of water in different tanks ranged between 27.17^{0}C -29.04^{0}C, 2.03-4.03 ppm and 7.19-7.93 respectively with little variation) under three different treatments (T_1, T_2, T_3) each of them have three replication (R_1, R_2, and R_3) and each tank contain 150 individuals (females : males) in each replication. The average total weight of males and female in each treatment was 47gm and 35gm respectively. The females under treatment T_1, T_2 and T_3 were treated with carp PG extract (prepared by homogenization of PG with a small volume of distilled water and that was carefully transferred to a centrifuge tube by using distilled water and centrifuged for 5 min at 3000 rpm to ensure complete transfer using following formula amount of PG required was calculated, Weight of carp PG (mg) (wt) = W_b × P_t/100 ; Where W_b represents total of the body weight of all the fishes injected and P_t represents the rate in mg of carp PG injected/kg body weight under a particular treatment) at the doses of 17, 15 and 13 mg/kg body weight respectively whereas males were treated at the doses of 14, 12 and 10 kg/body weight (Table 3) and through a 1ml hypodermic syringe the freshly prepared solution was injected intramuscularly to the fish on the dorsal side above the lateral line. The dose was divided into two volumes (40 % & 60 %) and injected to the broods with 6 hours interval. During injection needle was inserted at about 45° angles.

Table 1. Collection sites of fish samples (*M. vittatus*) from different stocks in Bangladesh

SI. No.	Date	Sources	Stocks	No. of Individuals
1.	12-02-2011	Brahmaputra river, Sutiakhali, Mymensingh	Brahmaputra river	750
2.	18-12-2011	Chamtaghat Kishorganj	Kishorganj Haor	550
3.	29-12-2011	Brahmaputra river, Somvhogongh, Mymensingh	Brahmaputra river	700
4.	27-02-2012	Chamtaghat Kishorganj	Kishorganj Haor	450
5.	28-02-2012	Brahmaputra river, Somvhogongh, Boraikandi	Brahmaputra river	115

Table 2. Composition of experimental feed ingredients

Ingredients	Inclusion level (%)	Preparation of 200 g	Preparation of 500 g
Wheat flour	10	20	50
Wheat bran	15	30	75
Rice bran	20	40	100
Maize meal	13.5	27	67.5
Fish meal	40	80	200
Vitamin-B	0.5	1	2.5
Vitamin-E	2.5 IU/g	500 IU	1250 IU

Table 3. Lay out of the experiment for PG doses in three treatments

Treatments	Replication	Stocking density(no./Tanks)	Gender	Dose(mg/ kg/body weight)		
				1st	Interval	2nd
	R_1	150	Female	3	6h	14
T_1	R_2	150				
	R_3	150	Male	6		8
	R_1	150	Female	2	6h	13
T_2	R_2	150				
	R_3	150	Male	5		7
	R_1	150	Female	1	6h	12
T_3	R_2	150				
	R_3	150	Male	4		6

Table 4. Average fertilization, hatching and survival rates of *Mystus vittatus* at different treatment

Treatment	Fertilization rate (Ave± Sd)	Hatching rate (Ave± Sd)	Survival rate (Ave± Sd)
T1	67±17.77	55±4.58	4±3.60
T2	74±7.93	67.66±7.23	4.66±2.51
T3	91.33±2.08	85±2	8±1.00

Table 5. The physico-chemical conditions of water in experimental bowls under different c with different PG doses

Treatment	Parameters	Initial sampling	1st sampling	2nd sampling	3rd sampling
T_1	Average temp. °C	28.00±0.5	27.82 ± 0.21	27.17± 0.05	28.87± 0.25
	Average DO ppm	3.54± 0.08	3.89 ± 0.27	4.03 ± 0.14	3.93 ± 0.14
	Average pH	7.47± 0.19	7.94 ± 0.23	7.38 ± 0.07	7.19 ± 0.09
T_2	Average temp. °C	28.30±0.13	27.54 ± 0.07	27.18± 0.08	27.6 ± 0.17
	Average DO ppm	3.47 ± 0.12	3.81 ± 0.24	3.73 ± 0.15	3.25 ± 0.16
	Average pH	7.59 ± 0.19	7.70 ± 0.10	7.45 ± 0.23	7.66 ± 0.09
T_3	Average temp. °C	27.86 ±024	28.05 ± 0.08	27.45± 0.22	28.4 ± 0.15
	Average DO ppm	3.59 ± 0.04	3.75 ± 0.61	3.42 ± 0.73	2.55 ± 0.46
	Average pH	7.32 ± 0.28	7.42 ± 0.07	7.44 ± 0.18	7.53 ± 0.11

Determination of Ovulation, fertilization and hatching rate

For determination of fertilization and hatching rates of fertilized eggs produced by each treatment, a portion of eggs from each female was taken separately and incubated in bowls of 15 liters. Soon after fertilization, the embryonic development started and the fertilized eggs looked watery and slightly transparent. Within 1 h of incubation, the numbers of fertilized and unfertilized eggs from each bowl were counted based on the color of the eggs. The unfertilized eggs turned opaque and whitish in color. No. of all fertilized eggs was counted contained in the bowls. After completion of hatching, the number of larvae from each bowl was counted by siphoning them out. Percent ovulation fertilization and hatching rates were calculated using following formulae:

$$\% \text{ ovulation} = \frac{No.\ of\ fish\ ovulated}{Total\ no.\ of\ fish\ injected} \times 100$$

$$\% \text{fertilization} = \frac{No.\ of\ fertilized\ eggs}{Total\ no.\ of\ eggs\ (fertilized+unfertilized)} \times 100$$

$$\% \text{ hatching} = \frac{No.\ of\ eggs\ hatched}{Total\ no.\ of\ eggs\ (fertilized+unfertilized)} \times 100$$

Observation of the embryonic and larval development

After the egg samples were collected randomly from the bowls with the help of a dropper and were taken in a petridish containing water for studying the embryonic developmental stages of *M. cavasius* at every 15 min, 30 min and 1 h interval till completion of morula, gastrula and hatching stage respectively. Then larvae samples were collected from the incubator. Initially samples were collected at daily intervals. At least 10 eggs and larvae undergoing embryonic and larval developmental process were observed by microscope (Optica C×41) and digital camera together with software (Magnus MIPS- Microsoft Image Processing System) for embryonic and larval developmental to obtain precise information about developmental stages.

First feeding

Although the hatchlings of *M. cavasius* get nutrition from the yolk sac upto 3 days after hatching, the larvae were provided first feeding from 3rd days (approximately 70h) after hatching at ambient temperature of 27-29°C. Hard boiled chicken egg yolk was provided as first feed for the hatchlings upto satiation level. Three days after fertilization live zooplankton (Tubified worms) were supplied as food. The larvae reared up to 21 days and then transferred to nursery pond for further rearing.

Statistical analysis

For statistical analysis of data, a one-way analysis of variance (ANOVA) was followed. Significant results were further tested by using Tukey's Multiple Comparison test to identify significant difference among the means. The statistical data analysis was carried out with the aid of the computer software SPSS version 17 (SPSS, 1999).

RESULTS

Ovulation rate

Females treated with three different doses of PG extract showed no difference in the effectiveness of the doses on including ovulation in females. All of treated females were ovulated.

Fertilization rate

Fertilization rates of ovulated eggs in three different treatments (T_1, T_2, and T_3) showed marked difference in the effectiveness among three doses of PG extracts (Table 3). Fertilization rates of eggs were obtained from females treated with treatment T_1, T_2 and T_3 showed 67.00 ± 17.78, 74.00 ± 7.94 and $91.33\pm2.08\%$ fertilization, respectively. The highest fertilization rate (91.33%) was recorded in T_3 (13 mg PG/kg/bw in female, 10 mg PG/kg/bw in male) whereas the lowest fertilization rate (67%) was found in T_1 (17 mg PG/kg/bw in female, 14 mg PG/kg/bw in male). Duncan's New Multiple Test indicates that T_1 was significantly ($P<0.05$) lower than T_3 and T_2 but there was no significant difference between T_3 and T_2. (Table 4)

Hatching rate

There was marked difference in the hatching rates of fertilized eggs in three different treatments (T_1, T_2 and T_3) (Table 4). Hatching rates of fertilized eggs obtained from females treated with treatment T_1, T_2 and T_3 were $55.00\pm4.58\%$, $67.67\pm7.23\%$ and $85\pm2\%$, respectively. Duncan's New Multiple Test for hatching rate showed that T_3 was significantly ($P<0.05$) higher than T_3 (13 mg PG/kg/bw in female, 10 mg PG/kg/bw in male) was significantly ($P<0.05$) higher than T_1 (17 mg PG/kg/bw in female, 14 mg PG/kg/bw in male) and T_2 (15 mg PG/kg/bw in female, 12 mg PG/kg/bw in male) but there was no significant difference between T_1 and T_2.

Survival rate

The survival rate was found to be $4.00\pm3.6\%$, $4.66\pm2.52\%$ and $8.00\pm1\%$ in treatment T_1, T_2 and T_3, respectively after 21 days of experimental period and (Figure 5). Duncan's New Multiple Test revealed a significantly ($P<0.05$) higher survival rate in T_3 (15 mg PG/kg/bw in female, 12 mg PG/kg/bw in male) than T_2 and T_1 but there was no significant difference between T_2 and T_1.

Physico-chemical condition of water

The physico-chemical conditions of water in experimental bowls under different treatments with different PG doses are shown in Table 5. Temperature, DO and pH of water in different bowls ranged between 27.17^0C -29.04^0C, 2.03-4.03 ppm and 7.19-7.93, respectively with little variation.

Observation of the embryonic and larval development of *M. vittatus*

Changes in the pattern of the entire structure of an organ or of a specific organ in relation to the environment are decisive for evaluating the developmental patterns of a species (Balo, 1999). Changes in structure emphasize the thresholds between embryonic, larval, and post-larval development from the onset of cleavage or epiboly, or at the time of organogenesis, respectively (Kovac, 2000; Carlos et al., 2002).

DISCUSSION

Optimization of the dose of PG for induced breeding of *M. vittatus*

Dose optimization is an important aspect for successful breeding programme. To standardize the dose of PG for successful ovulation, many scientists attempted to conduct experiments [9] as the catfishes do not spawn in the laboratory condition but readily respond to injection of fish and frog pituitary gland extract and to mammalian gonadotropins (Haniffa and Sridhar, 2002) but there are remains ambiguity among the doses reported by various workers. At doses combination 6-12 mg PG/kg body weight, females respond to ovulation   for at 1:2 male and female ratios where male was treated 3-6 mg PG/kg body weight revealed 80% fertilization and 56% hatching rates. Mean survival percentage of the spawns up to 10 days was 60% [11] which is slight a bit lower than the present experiment in terms of doses, fertilization (91.33±2.08%), hatching (85.00±2%) rates but opposite is true for survival rate (8.00±1%)   as the days pass by this may due to temperature, that is optimum (24-30°C), and causes the hatching rate increased and ranging from 48.0±0.118, 74.33±0.232 in *Cyprinus carpio* (El-Gamal, 2009) neutral pH as reported (Nchedo and Chijioke, 2012) and other factors such as (a) age and physical state of fish, (b) the seasonal variation, (c) environmental parameters such as water temperature, dissolved oxygen etc. (d) source of fish (wild or farmed) and most importantly (e) the source, age and maturity of the donor of PG used in the experiment. Whereas (Mijkherjee et al., 2002) [14] with different doses (1-2.5 ml/kg body weight) of PG at 1:2 females and males ratios of catfishes (Pabda, *Ompok pabda* and Tengra *Mystus guillo*) and stated that 2.5 ml/kg body weight of female showed maximum ovulation, hatching (80%) rate but all the females did not show 100% ovulation may be due to species difference.

Observation of the embryonic development of *M. vittatus*

Stage	Phase	Time after fertilization	Developmental landmarks	Fig. No.
I	Unfertilized egg	00 min	Opaque, demarsal, spherical and whitish in colour	1(a)
II	Fertilized egg	00 min	Transparent, demarsal, spherical and brownish in colour	1(b)
III	Cell division	35-40 min	Cleavage	1(c)
IV	Morula	2h	Cleavage resulted into 64 cells and were arranged in 3-4 layers	1(d)
V	Blastula	2h 35min	Spherical shape, flat border between blastodisc and yolk	1(e)
VI	Gastrula	2h 55min	Cleavage resulted into 64 cells and were arranged in 3-4 layers	1(f)
VII	Head and tail bud formation	9h 10	Head and tail rudiment visible. Notochord became visible, auditory and optic bud developed	1(g)
VIII	Just before hatching	23h 30min	Embryo encircled the whole yolk. The olfactory pits and auditory vesicles were prominently visible. Melanin pigmentation was developed. Continuously beat the egg shell by the caudal region especially around the middle part of the body	1(h)
IX	Newly hatched larvae	24 h	Slender, transparent and the yolk sac oval in shape. Anus situated at almost mid ventrally. The length of newly hatched larvae about 3-4 mm	1(i)

Observation of the larval development of *M. vittatus*

Stage	Age of larvae	Characteristics	Plate No.
I	1h 55 min	Mouth was not yet developed. Heart became more distinct. Larvae tried to move by propelling the tail.	2 (a)
II	5h	Melanophores appeared on the head, around the yolk sac. The anterior part began to thicken and stronger.	2 (b)
III	9 h	A tubular pulsating heart appeared. Eye and anus slightly visible.	2 (c)
IV	18h	Eye spot with a dark pigmented area and barbells were found in the forms of tiny knobs. Pectoral fin buds were seen. Prominent chromatophore was present on the head region. Larvae swam haphazardly.	2 (c)
V	26h	Distinct heart functioned actively; reddish blood was seen around the heart region. Mouth was formed as a small opening and the anal pore also opened.	2 (c)
VI	30h	Pectoral fin rudiment faintly visible.	2 (d)
VII	48h	Burbles appeared. Brain lobe clearly distinguished. The heart functioned actively.	2 (e)
VIII	72h	Yolk sac completely disappeared and larvae started feeding	2 (f)

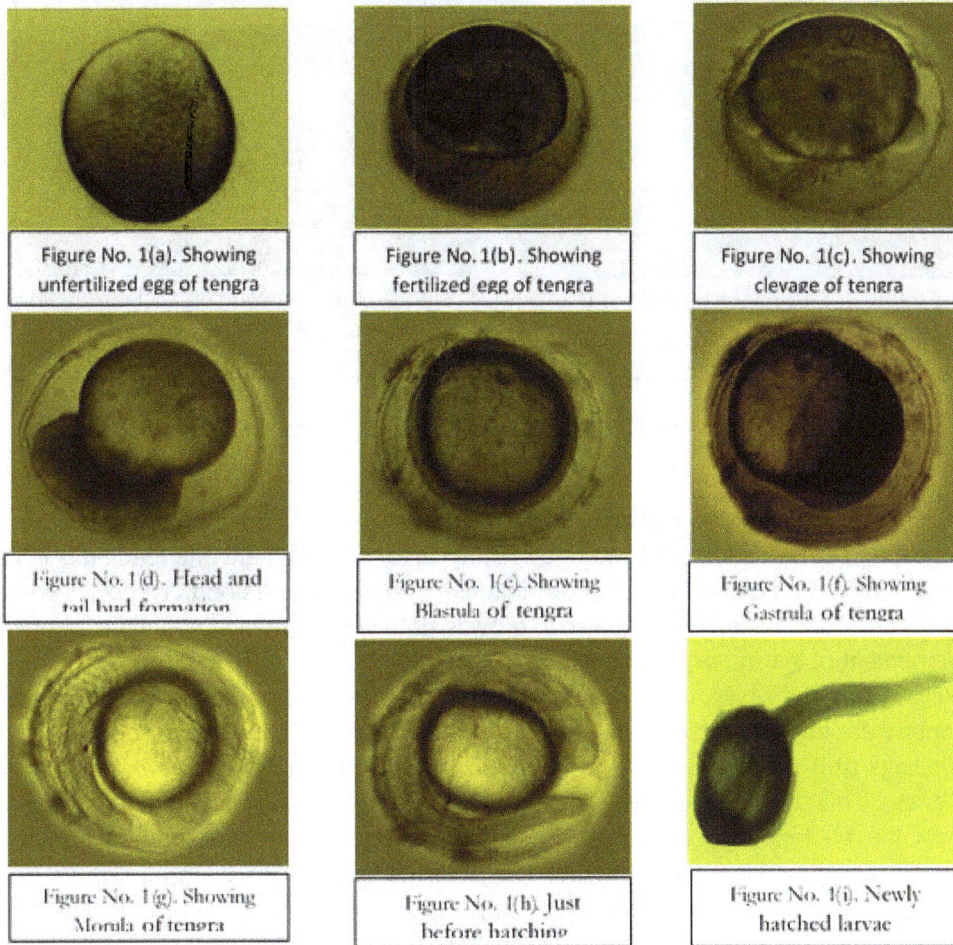

Figure No. 1(a). Showing unfertilized egg of tengra

Figure No. 1(b). Showing fertilized egg of tengra

Figure No. 1(c). Showing clevage of tengra

Figure No. 1(d). Head and tail bud formation

Figure No. 1(e). Showing Blastula of tengra

Figure No. 1(f). Showing Gastrula of tengra

Figure No. 1(g). Showing Morula of tengra

Figure No. 1(h). Just before hatching

Figure No. 1(i). Newly hatched larvae

Figure 1. Stages of the embryonic development of *M. vittatus*

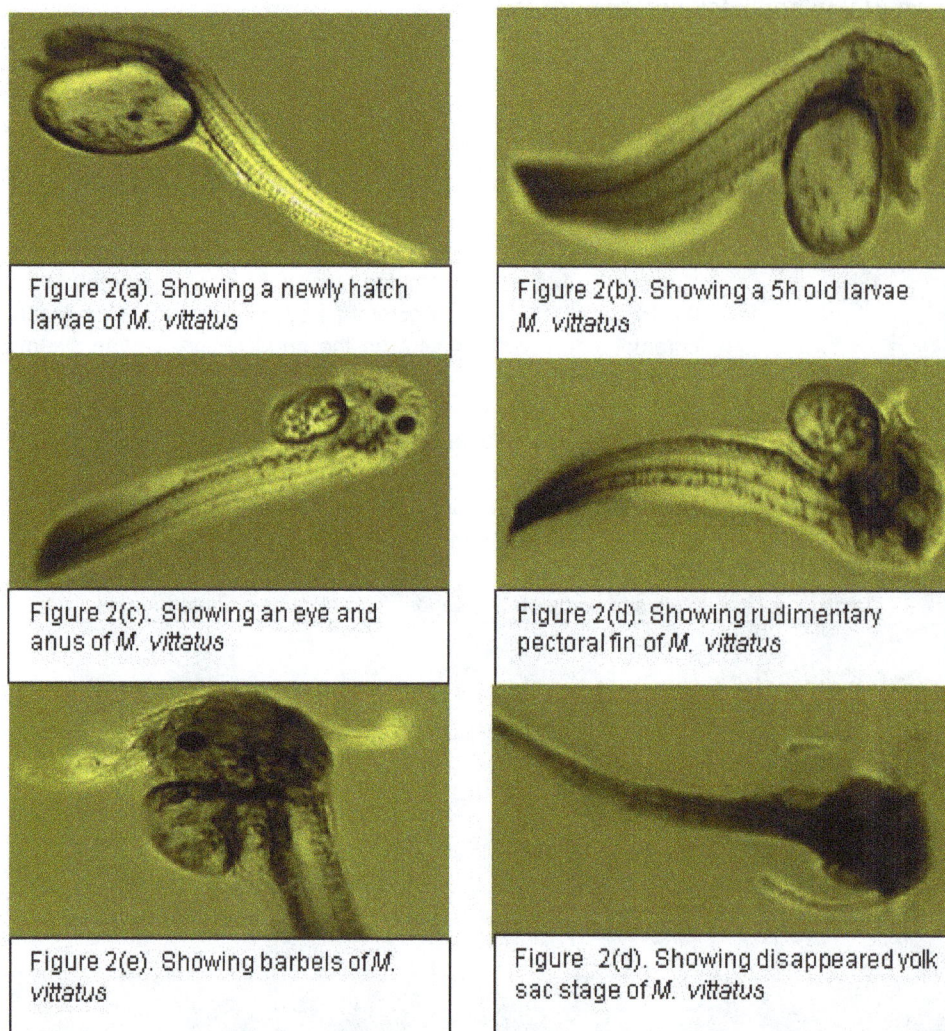

Figure 2(a). Showing a newly hatch larvae of *M. vittatus*

Figure 2(b). Showing a 5h old larvae *M. vittatus*

Figure 2(c). Showing an eye and anus of *M. vittatus*

Figure 2(d). Showing rudimentary pectoral fin of *M. vittatus*

Figure 2(e). Showing barbels of *M. vittatus*

Figure 2(d). Showing disappeared yolk sac stage of *M. vittatus*

Figure 2. Stages of the embryonic development of *M. vittatus*

Observation of the embryonic and larval development

To expand catfish culture, knowledge of early larval development and feeding is imperative. But the embryonic and larval development of this fish is poorly understood. Therefore the present study was conducted to investigate and also to provide detailed information about the embryonic and larval development of this important fish species.

The unfertilized eggs of *M. vittatus* were adhesive, whitish in colour and slightly smaller while the fertilized ones were transparent, demersal, spherical, adhesive and brownish in colour which agreed with the findings of (Puvaneswari et al., 2009) except in the reported colour which they found as brownish green. This difference in colouration may be due to the species difference or may be due to colour of the background container. The fertilized eggs of *M. vittatus* were adhesive. The adhesiveness of eggs is the special character of other catfish species such as *Clarias gariepinus* (Osman et al., 2008), *Mystus montanus* (Arockiaraj et al., 2003) and *Pangasius sutchi* (Islam, 2005).

In the present study, the diameter of the fertilized eggs ranged between 1.0 and 1.3 mm. Variation in egg size was also recorded for the eggs of *Clarias gariepinus* (Osman et al, 2008). This variation might be attributed to the species variation and brood size. In this study, first cleavage took place within 25-30 min post-fertilization at the water temperature of 27-28°C but in *Clarias gariepinus* and *Mystus cavasius* first cleavage took place within 40-50 min post-fertilization reported from (Khan and Mollah, 1998; Rahman et al. 2004) at 28.5 and 26°C respectively. This variation might be due to species difference and other environmental factors. In the present observation, morula stage reached within 2.40 h post-fertilization. Gastrula stage was found in *M.vittatus* at 7.30 to 9.30 h of fertilization at 27-29°C. Previous report from (Puvaneswari et al., 2009) in *Heteropnuestes fossilis* also indicated the duration of 7 h to reach the gastrula stage. Just 1-2 h before hatching, the embryo of *M. vittatus* showed twisting movements inside the egg capsule. The similar hatching behaviour was found in different fishes reported by (Puvaneswari et al. 2009; Osman et al., 2008).

In the present study, hatching commenced from 24 h at 28.87±0.25°C which is similar to other findings reported from (Arockiaraj et al., 2003). In *Clarias gariepinus* hatching started after 18 h at 28.5°C and was completed within 22 h which is lower than the present observation (Khan et al, 1998). The development of embryo and the variability of hatching time in fertilized eggs of most of the fish are generally influenced by the temperature of water (Mollah and Tan, 1982). At lower temperature the hatching started late and the duration of hatching was longer. In *Clarias macrocephalus* hatching started within 22 h at 30°C and 34 h was required to hatch at 25°C. But at 20°C no hatching was observed (De Graaf and Janssen, 1996). Length of the newly hatched larvae of this species was around 1.0 mm which is in the ranges of the findings of (Arockiaraj et al., 2003). Variation in length of the newly hatched larvae was recorded by several scientists. Ogunji and Rahe (1999) recorded the length of newly hatched larvae of *H. longifilis* vary from 4.09 to 4.9 mm. These variations may be related to the size of the eggs. According to (Bagarinao and Chua, 1986), egg diameter is positively correlated with larval length and weight at hatching. In the present study, larvae of *M. vittatus* started to feed at 48 h after hatching which is similar to other previously reported studies stated by (Puvaneswari et al., 2009).

The present work generated some information on induced breeding, embryonic and larval development of *M. vittatus*. The study consisted of two experiments. The first experiment dealt with the optimization of the doses of carpPG extract on breeding performance of *M. vittatus*. The second one was conducted to know embryonic and larval development of the same species.

CONCLUSION

It is however, difficult to pinpoint the reason for such differing results because a number of factors affect the biological experiment particularly involving hormones. Several major factors that may have bearing on the result are: But upon all consideration a PG dose of 13 mg/kg may be recommended for this species.

REFERENCES

1. IUCN, Bangladesh, 2000. List of threatened animals of Bangladesh. Paper presented in the special workshop on Bangladesh Redbook of threatened animals, 22 February 2000, Dhaka. pp. 13.
2. Perennou C and V Santharam, 1990. An anthropological survey of some wetlands in south-east India. Journal of Bombay National Historical Society, 87: 354-63
3. Roos N, MM Islam and SH Thilsted, 2003. Small indigenous fish species in Bangladesh: contribution to vitamin a, calcium and iron intakes. Journal of Nutrition, 133 4021S-4026S.
4. SPSS, 1999. Software Program of Statistical Analysis, Version 17.0 Edition for Windows. SPSS Incorporation Chicago IL US.
5. Balo K, 1999. Alternative ways how to become a definitive phenotype or a juvenile (and on some persisting linguistic offences). Environmental Biological Fisheries, 56: 17-38.

6. Kovac V, 2000. Early development of Zingel streber. Journal Fisheries Biology, 57: 1381-1403.

7. Carlos AM, MC Sanchez, GS Papp, AD Parra and LG Ross, 2002. Observations on spawning, early development and growth of the puffer fish *Sphoeroides annulatus*. Journal of Aquaculture Tropical, 17: 59–66.

8. Thakur NK and P Das, 1974. Length weight relationship of Shinghi, *Heteropneustes fossilis*. Journal of Inland Fish Society Indian, 6: 95-96.

9. Haniffa MAK and S Sridhar, 2002. Induced spawning of spotted murrel (*Channa punctatus*) and catfish (*Heteropneustes fossilis*) using human chorionic gonadtropin and synthetic hormone (ovaprim). Veterinary Archive, 72: 51-56.

10. Rahi Md, SK Lifat, I Shahinur and MS Shah, 2011. Study of fecundity and induced breeding of *Mystus vittatus*. Bangladesh Journal of Zoology, 39: 205-212.

11. Abd El-Hakim and E El-Gamal, 2009. Effect of Temperature on Hatching and Larval Development and Mucin Secretion in Common Carp, *Cyprinus carpio* (Linnaeus, 1758). Global Veterinaria, 3: 80-90.

12. Nchedo AC and OG Chijioke, 2012. Effect of pH on Hatching Success and Larval Survival of African Catfish (Clarias gariepinus). Nature and Science, 10: 21-25.

13. Mijkherjee M, A Praharaj and S Das, 2002. Conservation of fish stocks through artificial propagation and larval rearing technique in West Bengal, India. Aquaculture Asia, 7: 8-11.

14. Puvaneswari S, K Marimuthu, R Karuppasamy, MA Haniffa, 2009. Early embryonic and larval development of Indian catfish, *Heteropneustes fossilis*. EuroAsia Journal of Biological Science, 3: 84-96.

15. Osman AGM, S Wiertz, IA Mekkawy, J Verreth and F Kirschbaum, 2008. Early development of the African catfish *Clarias gariepinus* focusing on the ontogeny of selected organs. Journal of Applied Ichthyology, 24: 187-195.

16. Arockiaraj J, MA Haniffa, S Seetharaman and SP Singh, 2003. Early development of a threatened freshwater catfish *Mystus montanus* (Jerdon). Acta Zoologica Taiwanica, 14: 23-32.

17. Islam M, 2005. Embryonic and larval development of Thai pangas (*Pangasius sutchi* Fowler, 1937). Development and Growth Diffraction, 47: 1-6.

18. Khan MMR and MFA Mollah, 1998. Embryonic and larval development of African catfish *Clarias gariepinus*. Bangladesh Journal of fisheries, 21: 91-97

19. Rahman MR, MA Rahman, MN Khan and MG Hussain, 2004. Observation of the embryonic and larval development of silurid catfish, Gulsha (*Mystus cavasius* Ham.) Pakistan Journal of Biological Science, 7: 1070-1075.

20. Mollah MFA and ESP Tan, 1982. Effect of incubation temperature on the hatching catfish (*Clarias macrocephalus*) eggs with an illustration on the larval stages. Malaysian National Journal, 36: 123-131.

21. De Graaf GJ and H Janssen, 1996. Artificial reproduction and pond rearing of African catfish (*Clarias gariepinus*) in sub Sahran Africa. FAO Fisheries Technical paper, 362: 1-73.

22. Ogunji JO and RE and Rahe, 1999. Larval development of the African catfish *Heterobranchus longifilis*. 1840 (Teleostei Claridae) and its larval behavior. Journal of Aquaculture Tropics, 14: 11-25.

23. Bagarinao T and TE Chua, 1986. Egg size and larval size among teleost simplications to survival potential. In: Maclean J, Dizon LB, Hossils LV (eds). The First Asain Fisheries Foroum, Asain Fisheries society, Manila. pp. 651-656.

2

AQUA-DRUGS AND CHEMICALS: IMPACT ON FISH HEALTH AND PRODUCTION IN MYMENSINGH, BANGLADESH

Gias Uddin Ahmed, Md. Towhid Hasan, Md. Ali Reza Faruk, Md. Khalilur Rahman[1] and Md. Nazmul Hoque

Department of Aquaculture, Faculty of Fisheries, Bangladesh Agricultural University, Mymensingh-2202, Bangladesh; [1]Bangladesh Fisheries Research Institute, Mymensingh-2202, Bangladesh

*Corresponding author: Gias Uddin Ahmed; E-mail: giasa50@gmail.com

---

ARTICLE INFO	ABSTRACT

Key words
Aqua-drugs
Chemicals
Fish
Production
Disease
Histology

The present study was conducted to evaluate the impacts of aqua-drugs and chemicals on fish health and production in Mymensingh, Bangladesh from July 2013 to June 2014. Data were collected through personal contact, market survey, focus group discussion and participatory rural appraisal with fish farmers, associations and aqua-drug sellers. Fifty five different types of aqua-drugs and chemicals were recorded in the study area, among those, 20 types were widely used by the farmers for different fish disease treatment. It was recorded that renamycine, cotrim vet, ossi-c, polgard plus and timsen were used for the treatment of EUS in pangus, tilapia and koi which had an average recovery of 75-85%. For the treatment of edwardsiellosis in pangus and koi, farmers used potassium permanganate, timsen, polgard plus, geolite gold and renamycine which had an average recovery of 65-80%, and for the treatment of dropsy in tilapia, aquamycine, ossi-c and potassium permanganate were used which had an average of 80-85% recovery. Histopathology of gill and liver of fishes were almost normal in control ponds, whereas, in drugs treated ponds the organs had pathological changes like necrosis, pyknotic cells, hemorrhage, hypertrophy, lamellar missing, talengiactasis and vacuums. However, production of pangus, tilapia and koi was 12000 kg/acre, 15000 kg/acre and 14000 kg/acre in drug treated ponds respectively, whereas, it was 5000 kg/acre, 9000 kg/acre and 8000 kg/acre in non-treaded ponds respectively.

INTRODUCTION

Over the last decade the rapid expansion of fish culture has been drawn an outstanding development in Bangladesh fisheries sector and it contributes 4.43% to the Gross Domestic Product (DoF, 2014). Aquaculture expands through the practice of improved extensive and semi intensive fish culture systems. For the successful aquaculture, technology is most needed (Subasinghe et al., 1996) as well as the application of different aqua-drugs and chemicals which enhance the production and disease resistance capacity. There is a long history behind the using of drugs and chemicals in aquaculture. A variety of aqua-drugs and chemicals are used in both inland and coastal aquaculture. The purposes of using chemicals and antibiotics are to improve health condition of aquatic animal, growth promotion (Ahmed et al., 2014), feed formulation, manipulation of production, transportation of live fish, pond construction, and overall the management of natural pond environment and water quality (GESAMP, 1997; Faruk et al., 2004 and Khan et al., 2011).

In past farmers used only some traditional chemicals like lime, salt, potassium permanganate, copper sulphate, formalin and bleaching powder (Hasan and Ahmed, 2002 and Plumb, 1992) but in recent years several pharmaceutical companies play a vital role to produce various types of commercial aqua-drugs and chemicals (Faruk et al., 2008). For health management of fish several types of antibiotics are used by farmers. The antibiotics, which have been applied in aquaculture for over fifty years for treating bacterial infections in fish (Shamsuzzaman and Biswas, 2012). The common ingredients of antibiotics are oxytetracycline, chlorotetracycline, amoxicilin, co-trimoxazoie, sulphadiazine and sulphamethoxozole (Plumb, 1992). Some common chemicals are used for health management including sodium chloride, formalin, malachite green, methyl blue, potassium permanganate and hydrogen per-oxide (Plumb, 1992). Potassium permanganate is the most widely used chemical for treating external protozoa and external bacterial infection. For treating fungal infection, external parasite on fish and fish eggs as flush, prolonged or indefinite treatment or fungal control sodium chloride and formalin is an old treatment used by the farmers (Plumb, 1992). Thus, present study was carried out to evaluate the impact of aqua-drugs and chemicals on fish health and production in inland aquaculture of Bangladesh.

MATERIALS AND METHODS

The present study was carried out in Trishal and Bhaluka upazillas in Mymensingh district from July 2013 to June 2014. Data were collected through questionnaire interview, personal contact, participatory rural appraisal (PRA) and focus group discussion (FGD) with fish farmers and associations, market survey and retailers of aqua-drugs and chemicals. The sample size varied from different target groups such as 12 to 15 farmers, 3 to 4 drug sellers or drug shops and 1 to 2 farmers association from each sampling stations.

The impact of different aqua-drugs and chemicals on fish health and production was measured through the farmer's opinions. Fish production was compared between culture systems using aqua-drugs and chemicals and without chemicals.

The samples were collected from the field level for health check through histological observation. Fish samples were collected from gill and liver. Sampling was done by a sharp scalpel and forceps and fixed in 10% neutral buffer formalin and kept in transparent plastic vials. Fish samples were processed in an automatic tissue processor (SHANDON, CITADEL 1000), embedded, sectioned using a microtome (Lecia JUNG RM 2035), stained with haematoxyline and eosin, mounted with Canada balsam and the slides were examined under a compound microscope (OLYMPUS, Model CHS, Japan). Then photomicrographs were taken by a photographic camera in Fish Disease Laboratory at BAU, Mymensingh.

RESULTS AND DISCUSSION

During the present investigation, seven categories and 34 pharmaceutical companies were recorded in the study area. From the research findings of Faruk et al. (2008) 33 pharmaceutical companies were found either producing or marketing aqua-drugs in Mymensingh district. Farmers used chemicals which were categorized as pond preparatory and water quality maintenance, oxygen supplementary, gas removal, growth promoters, disinfectants, antibiotics and disease treatment. According to Faruk *et al.* (2008) farmers of Mymensingh regions used different types of aqua-drugs and chemicals for various purposes like pond preparation, growth promotion, increasing oxygen concentration, disinfection, probiotics and disease treatment. In present study, 55 different types of aqua-drugs and chemicals were recorded in the study area. Among those, 20 types were widely used by the farmers for different disease treatment of fish such as bactitab, chlorsteclin, orgacycline, orgamycine-15%, oxy-d vet, oxysentin, renamox, renamycin, malachite green, methylene blue, bleaching powder, potassium permanganate, eco-solution, basudin, timsen, oxytetracycline, lime, formalin, oxolinic acid and sarafloxacin. From the research findings of Ahmed *et al.* (2014) in Mymensingh district farmers used 50 different types of aqua-drugs and chemicals on various purposes among those, 15 types of antibiotics and drugs were used by the farmers for the treatment of different fish diseases.

Impact on fish health and disease

In case of inland aquaculture of Mymensingh region various types of fish diseases were detected. EUS, edwardsiellosis, dropsy, pop eye, white spots and fin root diseases were found in pangus, tilapia and koi. From the research findings of Ahmed *et al.* (2014) in Mymensingh region EUS, dropsy and edwardsiellosis were observed in pangus, koi and tilapia. From the present study in Trishal upazilla farmers used zeolite, gastab, timsen, renamycin and polgard plus for the treatment of EUS in pangus, which had an average of 75-80% recovery (Table 1). However, in Bhaluka upazilla farmers used potassium permanganate, renamycine, cyprocine and cotrimvet for the treatment of EUS in pangus, which had an average of 80-85% recovery. According to Rahman (2012) in case of EUS, farmers of Jamalpur used oxysentin 20%, aquamycin and acimox powder and achieved 90% recovery with tilapia, rui, catla and pangus. In EUS affected tilapia and koi, farmers of Trishal upazilla used renamycin, polgard plus, ossi-c and aquamycine which had an average recovery of 80-85%. However, for the treatment of EUS affected tilapia in Bhaluka upazilla farmers used renamycin and ossi-c which had an average of 70-80% recovery (Table 1). According to Ahmed *et al.* (2014) to treat EUS affected tilapia farmers of Fulpur upazilla used renamycin, polgard plus and ossi-c with a result of 80-95% recovery. Rahman (2011) mentioned that EUS affected tilapia were treated with renamycin, polgard plus and ossi-c and achieved 95% recovery.

In the present study, farmers of Trishal upazilla used renamycin, timsen, ossi-c and polgard plus for the treatment of edwardsiellosis in pangus and koi with a result of 75-80% and 65-70% recovery respectively (Table 1). From the research findings of Ahmed *et al.* (2014) in edwardsiellosis affected pangus, farmers used renamycin, polgard plus, timsen and ossi-c having 80% recovery. Whereas, in Bhaluka upazilla, for the treatment of edwardsiellosis affected pangus and koi farmers used renamycin, polgard plus, timsen, geolite gold and ossi-c which had an average 75-80% recovery (Table 1). Rahman (2011) mentioned that edwardsiellosis affected Thai pangus were treated with renamycin, timsen, polgard plus and ossi-c having 80% recovery.

For the treatment of pop eye, tail and fin rot and dropsy farmers of both upazillas used aquamycin, ossi-c, lime, salt and renamycin having 70-85% recovery (Table 1). According to Ahmed *et al.* (2014) in dropsy affected tilapia farmers of Fulpur upazilla used aquamycin and ossi-c with a result of 95% recovery. In the present study it was observed that various spots on skin and scale dropped in some parts of koi, farmers of both upazillas used lime, salt, aquamix and vitamix with a result of 70-80% recovery (Table 1) according to the research findings of Ahmed *et al.* (2014).

Histological observations

From the present investigation section of gill of tilapia from Trishal were seen normal in control ponds (Figure 1), except hypertrophy and some lamellar missing of gill of koi from Trishal (Figure 3) and some lamellar missing of gill of koi from Bhaluka (Figure 4) in control ponds, which were in accordance with the findings of Rahman (2012). According to Ahmed *et al.* (2012) section of gill had almost normal structure in control ponds. However, in treated ponds, gill of tilapia from Bhaluka had lamellar missing, necrosis and hemorrhage (Figure 2). Section of gill of pangus from Trishal, there were seen talengiactasis and lamellar missing (Figure 5), and section of gill from Bhaluka having clubbing, cyst, talengiactasis and hemorrhage in treated ponds (Figure 6). Ahmed *et al.* (2014) mentioned that in case of gills of aqua-drugs and chemical treated ponds exhibited pathological changes like hypertrophy, hemorrhage, missing of secondary gill lamellae, clubbing and necrosis.

Photomicrograph of liver of tilapia and pangus from Trishal were normal in control ponds (Figure 7 and Figure 11). Section of liver of koi and pangus from Trishal and Bhaluka were almost normal except some vacuums in control ponds (Figure 7, Figure 9, Figure 11 and Figure 12). According to Rahman (2012) liver of fishes were almost normal in control ponds. From the research findings of Ahmed *et al.* (2014) in control ponds section of fish liver were almost normal except some vacuums. However, in treated ponds, section of liver of tilapia from Bhaluka had vacuums, necrosis and pyknotic cells (Figure 8). Section of liver of koi from Bhaluka had vacuums and hemorrhage in treated ponds (Figure 10). Ahmed *et al.* (2014) reportated that some important pathological changes such as hemorrhage, necrotic hepatocytes, pyknotic cells and vacuums were recorded in the liver of chemical treated fishes. According to Rahman (2012) liver of chemical treated fish had some important pathological changes such as hemorrhage, necrosis, pyknotic cell and vacuums.

Table 1. Impact of aqua-drugs on fish health and disease in Mymensingh

Study areas	Species	Diseases	Drugs/chemicals with dose	Recovery (%)
Trishal	Pangus	EUS	Zeolite 200g/dec, Gastab 2-3g/dec , Timsen 0.6g/dec, Cotrimvet 2g/kg feed	75-80
		Edwardsiellosis	Renamycin 5g/kg feed, Timsen 80 gm/33 dec, Ossi-C 3 gm/kg feed, Polgard plus 5 ml/decimal	75-80
		Fin root	Lime 0.5-1 kg/dec, salt 0.5-1 kg/dec	60-65
	Tilapia	EUS	Renamycin 50mg/kg body weight, Polgard plus 500 ml/acre, Ossi-C 3 g/kg feed	80-85
		Dropsy	Aquamycine 1-2 g/feed, Ossi-C 3 g/kg feed	80-85
	Koi	Edwardsiellosis	Renamycin 5g/kg feed, Ossi-C 3 gm/kg , Polgard plus 5 ml/decimal	65-70
		EUS	Aquamycine 1-2 gm/feed, Ossi-C 3 g/kg feed, Polgard plus 5 ml/decimal	80-85
Bhaluka	Pangus	EUS	$KMnO_4$ 3kg/dec, Renamycine 5g/kg feed , Cotrimvet 2g/kg feed, Revoflavin 50 tab/kg feed, Tetravet 5g/kg feed, Fish curapus 20g/dec	80-85
		Edwardsiellosis	Renamycin 5g/kg feed, Ossi-C 3 g/kg feed, Polgard plus 5 ml/decimal, Geolite gold 200-250 g/decimal	75-80
		Fin rot	Lime 0.5-1 kg/dec, salt 0.5-1 kg/ dec	70-75
		Pop eye	Renamycine 5g/kg feed	70-75
		Fat deposition	Livabid 10ml/kg feed, Cholin chloride 10ml/kg feed	50-55
	Tilapia	EUS	Renamycin 50 mg/kg body weight, Ossi-C 3 g/kg feed	75-80
		White spot	Lime 0.5-1kg/dec, salt 0.5-1kg/dec, Aqua mix 5g/kg feed, Vita mix-F-Aqua5g/kg feed	75-80
	Koi	Edwardsiellosis	Renamycin 5g/kg feed, Ossi-C 3 g/kg , Polgard plus 5 ml/decimal	75-80
		White spot	Lime 0.5-1kg/dec, salt 0.5-1kg/dec, Aqua mix 5g/kg feed, Vita mix-F-Aqua5g/kg feed	70-75

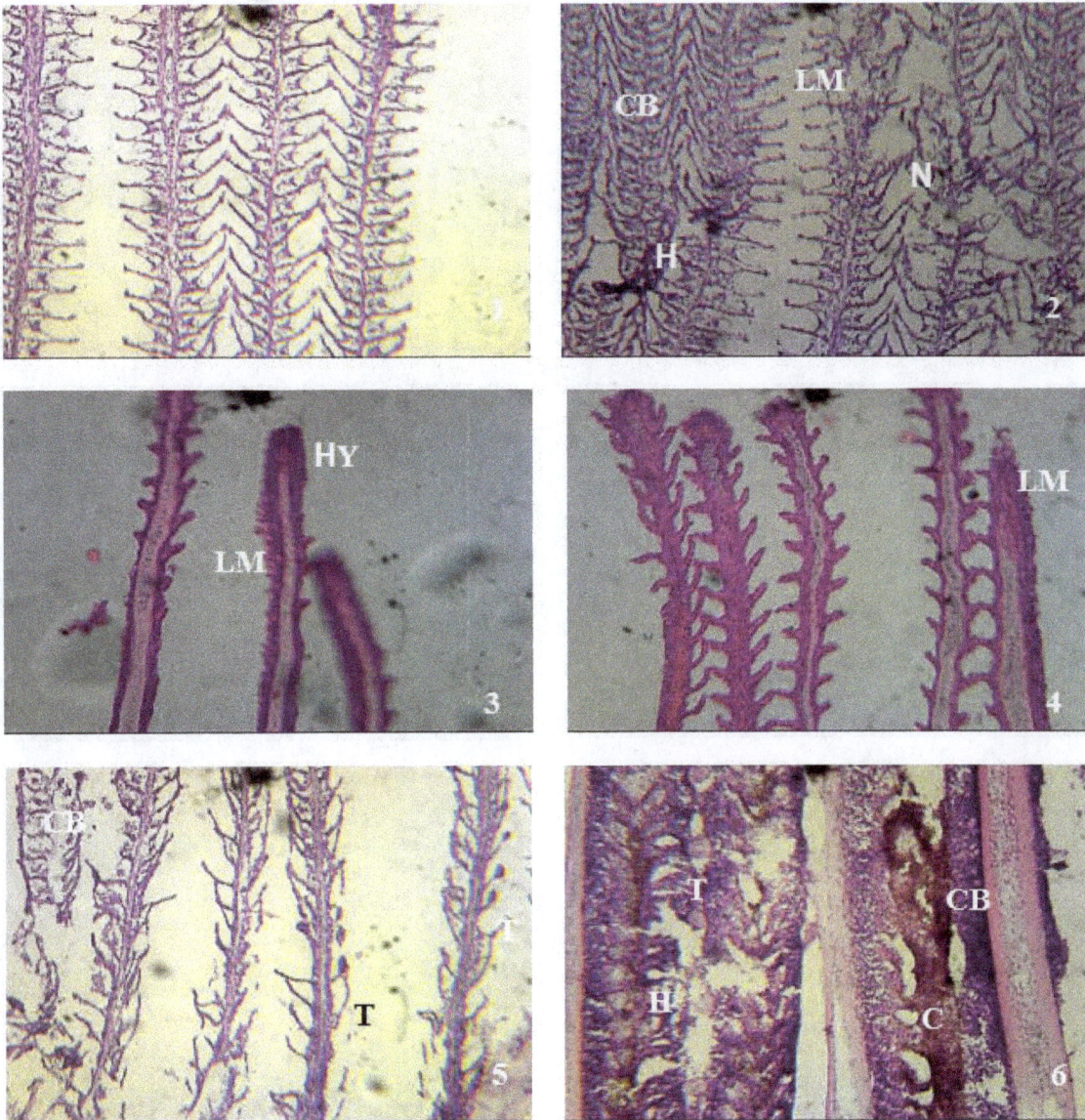

Figure 1. Cross section of normal gill of tilapia from Trishal from a control pond; **Figure 2.** Photomicrograph of gill of tilapia from Bhaluka having clubbing (CB), hemorrhage (H), lamellar missing (LM) and necrosis (N) from a treated pond; **Figure 3.** Section of gill of koi from Trishal having hypertrophy (HY) and lamellar missing (LM) from a treated pond; **Figure 4.** Cross section of almost normal gill of koi from Bhaluka except showing some lamellar missing (LM) from a control pond; **Figure 5.** Photomicrograph of gill of pangus from Trishal having clubbing (CB) and talengiactasis (T) from a treated pond; **Figure 6.** Section of gill of pangus from Bhaluka having clubbing (CB), talengiactasis (T), cyst (C) and hemorrhage (H) from a treated pond (All figures: H & E x 125).

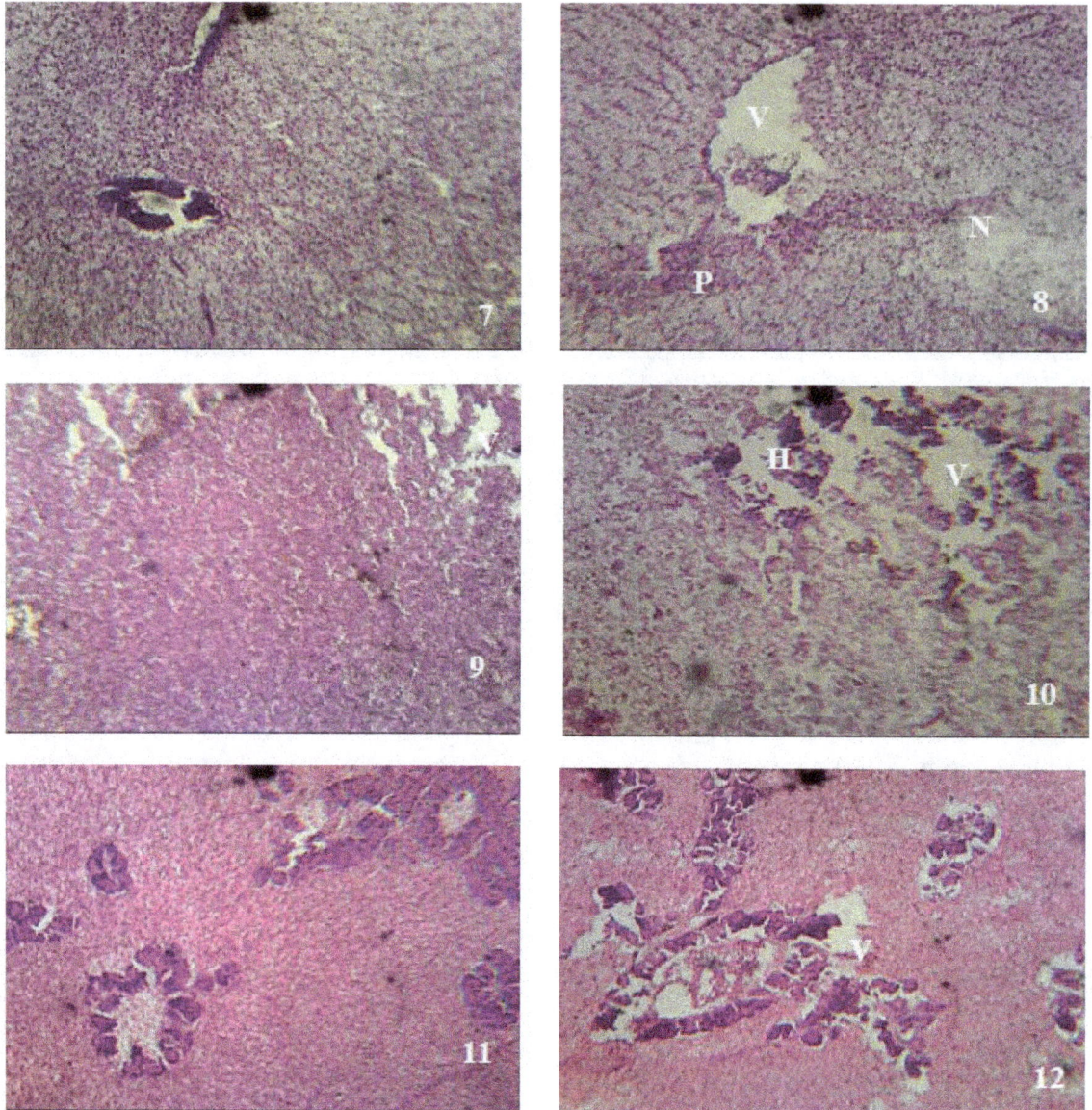

Figure 7. Photomicrograph of normal liver of tilapia from Trishal from a control pond; **Figure 8.** Cross section of liver of tilapia from Bhaluka having vacuum (V), necrosis (N) and pyknotic cell (P) from a treated pond; **Figure 9.** Section of an almost normal liver of koi from Trishal except showing vacuums (V) from a control pond; **Figure 10.** Photomicrograph of liver of koi from Bhaluka having vacuums (V) and hemorrhage (H) from a treated pond; **Figure 11.** Cross section of normal liver of pangus from Trishal from control pond; **Figure 12.** Section of almost normal liver of pangus except having vacuums (V) from a control pond from Bhaluka (All figures: H & E x 125).

Impact on fish production

In Mymensingh region, in Trishal upazilla Pangus production was 6000 kg/acre in control ponds, whereas, 12000 kg/acre in treated ponds. However, in Bhaluka upazilla pangus production was 5000 kg/acre in control ponds, whereas, 10000 kg/acre in treated ponds (Figure 13). From the research findings of Ahmed *et al.* (2012) in farmer's pond, production of Thai pangus in chemical treated ponds was higher 8100 kg/acre than in the non-treated ponds having 4800 kg/acre. Tilapia production was 9000 kg/ acre and 14000 kg/acre in control and treated ponds respectively in Trishal upazilla (Figure 13). However, in Bhaluka upazilla tilapia production was 10000 kg/acre and 15000 kg/acre in control and treated ponds, respectively. Koi production in Trishal upazilla was 9000 kg/acre and 14000 kg/acre in control and treated ponds respectively, however in Bhaluka upazilla 8000 kg/acre and 13000 kg/acre in control and treated ponds respectively (Figure 13). Shamsuddin (2012) mentioned that production of Thai pangus and Thai koi in Gouripur and Muktagacha Upazillas were almost double in the chemical treated ponds compared with non-treated ponds. According to the author, production of Thai pangus in BAU experimental ponds of control one was higher 7328.16 Kg/acre and in the treated one was 6400.08 Kg/acre (Figure 13).

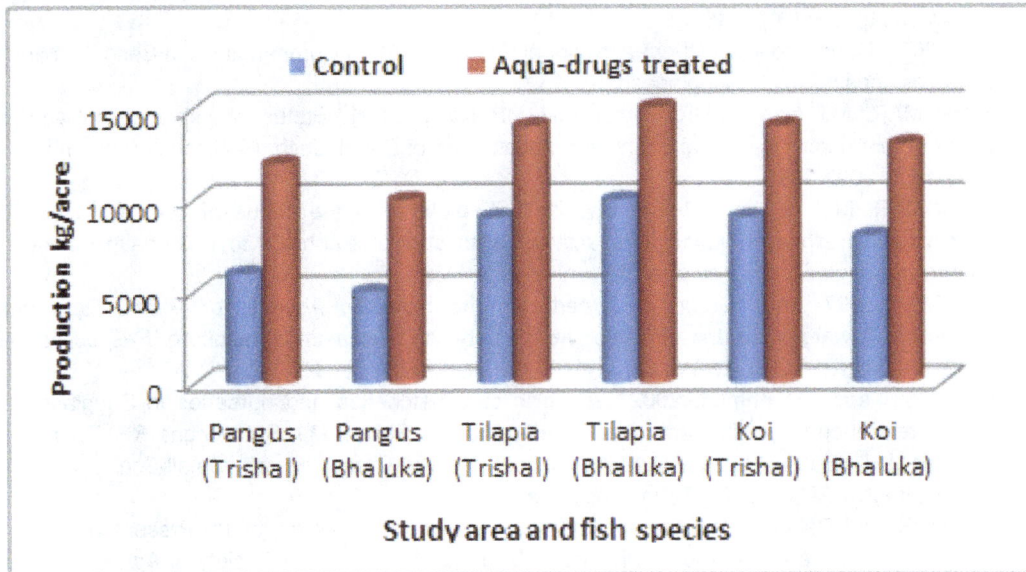

Figure 13. Fish production (kg/acre) in Mymensingh district.

CONCLUSION

Fifty five different types of aqua-drugs and chemicals were recorded in the study area, among those, 20 types were widely used by the farmers for different fish disease treatment. In Mymensingh region common fish diseases were recorded as EUS, fin rot, dropsy, white Spots, pop eye and edwardsiellosis. Potassium permanganate, renamycine, cyprocine and cotrimvet had an average recovery of 80-85% on EUS of pangus, whereas, renamycin, timsen, ossi-c and polgard plus had an average recovery of 75-80% on the treatment of Edwardsiellosis of pangus. From histological section of fish gill and liver, there were some pathological changes like necrosis, hemorrhage, pyknotic cell, lamellae missing, talengiactasis, clubbing and hypertrophy were observed in the above mentioned organs in drug treated ponds, whereas, some vacuums were observed in control ones. Production of pangus, was 12000 kg/acre, in drug treated ponds, whereas, 5000 kg/acre in control ponds. Aqua-drugs and chemicals had positive impacts on fish production and disease recovery, on the other hand, some remarkable pathological changes were observed in fish organs from drug treated ponds. So, the use of aqua-drugs and chemicals in ponds and ghers should be reduced in order to overcome adverse pathologies in fish organs.

ACKNOWLEDGEMENT

The authors gratefully acknowledge the co-operation of Bangladesh Fisheries Research Institute (BFRI) for granting financial support to carry out the present research.

REFERENCES

1. Ahmed GU, MAR Faruk and M Shamsuddin, 2012. Impact of aqua drugs and chemicals on health of fish. 5th Fisheries Conference & Research Fair 2012. Bangladesh Fisheries Research Forum (BFRF). pp. 39.

2. Ahmed GU, MAR Faruk, M Shamsuddin and MK Rahman, 2014. *Impact of aqua-drugs and chemicals on fish health.* In: MA Wahab, MA Shah, MS Hossain, MAR Barman and ME Haq (Editors). Advances of Fisheries Research in Bangladesh: 5th Fisheries Conference & Research Fair 2012. Bangladesh Agricultural Research Council, Dhaka, Bangladesh Fisheries Research Forum, Dhaka, Bangladesh. pp. 246.

3. DoF, 2014. Department of Fisheries, National Fish Week 2014 Compendium (In Bengali). Ministry of Fisheries and Livestock, Bangladesh. pp. 13.

4. Faruk MAR, MJ Alam, MMR Sarker and MB Kabir, 2004. Status of fish disease and health management practices in rural freshwater aquaculture of Bangladesh. Pakistan Journal of Biological Science, 7: 2092-2098.

5. Faruk MAR, MM Ali and ZP Patwary, 2008. Evaluation of the status of use of chemicals and antibiotics in freshwater aquaculture activities with special emphasis to fish health management. Journal of Bangladesh Agricultural University, 6: 381-390.

6. GESAMP, 1997. Joint Group of Experts on the Scientific Aspects of Marine Environmental Protection. Towards safe and effective use of chemicals in coastal aquaculture. *Respective Studies* (IMO/ FAO/ UNESCO/ IOC// WMO/ WHO/ IAEA/ UN/ UNEP, 65: 40.

7. Hasan MR and GU Ahmed, 2002. Issues in carp hatcheries and nurseries in Bangladesh, with special reference to health management. In: Arthur JR, Phillips MJ, Sabasinghe RP, Reantaso MB, MacRae LH (Eds.), Primary Aquatic Animal Health Carein Rural, Small-Scale. Aquaculture Development, FAO Fisheries Technology, 406: 147–164.

8. Khan MR, MM Rahman, M Shamsuddin, MR Islam and M Rahman, 2011. Present status of aqua drugs and chemicals in Mymensingh District. Journal of Bangladesh Society of Agricultural Science and Technology, 8: 169-174.

9. Plumb JA, 1992. Disease control in aquaculture. In: Shariff IM, Subasinghe RP, Arthur JR (Eds.) Disease in Asian Aquaculture. Fish health Section of the Asian Fisheries Society, Manila, Philippines. pp. 3–17.

10. Rahman H, 2012. Effects of aqua drugs on the health condition of farmed fishes of Jamalpur areas. MS Thesis, Department of Aquaculture, Bangladesh Agricultural University, Mymensingh, Bangladesh. pp. 31.

11. Rahman MM, 2011. Status and impact of commercial aqua drugs and chemicals on fish health at farmer's level. MS Thesis, Department of Aquaculture, Bangladesh Agricultural University, Mymensingh, Bangladesh. pp. 26.

12. Shamsuddin M, 2012. Impact of aqua-drugs and chemicals on health and production of fish. MS Thesis, Department of Aquaculture, Bangladesh Agricultural University, Mymensingh, Bangladesh. pp. 28.

13. Shamsuzzaman MM and TK Biswas, 2012. Aqua chemicals in shrimp farm: A study from south-west coast of Bangladesh. Egyptian Journal of Aquatic Research, 38: 275-285.

14. Subasinghe RP, U Barg and A Tacon, 1996. Chemicals in Asian aquaculture: need, usage, issues and challenges. *Use of Chemicals in Aquaculture in Asia*. In: JR Arthur, CR Lavilla-Pitogo, RP Subasinghe (Editors), Southeast Asian Fisheries Development Center, Aquaculture Department Tigbauan, Iloilo, Philippines. pp. 1-6.

3

COMPARATIVE STUDY ON GROWTH OF SUPERMALE TILAPIA AND MONOSEX TILAPIA IN EARTHEN MINI POND

Gias Uddin Ahmed, Habiba Aktar, Sudristi Chakma, Neaz Al Hasan and Mohammad Shamsuddin

Department of Aquaculture, Faculty of Fisheries, Bangladesh Agricultural University, Mymensingh-2202, Bangladesh

*Corresponding author: Gias Uddin Ahmed; E-mail: giasa50@gmail.com

ARTICLE INFO	ABSTRACT

Key words

Growth
Monosex
Supermale
Tilapia
Mini pond

An investigation was conducted to determine the comparative growth study of supermale tilapia and monosex tilapia in earthen mini ponds from May-July 2012. Four treatments were considered having two replicates. For supermale tilapia treatments were named as ST_1 and ST_2 and for monosex tilapia were MT_1 and MT_2. All the fish were of same age group having mean body weight of 1.4 g. Feeding frequency in all the treatments were two times a day. Fish were fed diet at a rate of 30% of their body weight for the first thirty days that was gradually reduced to 15% for the next thirty days and 5% till the termination of the experiment. Final weight, weight gain, average daily weight gain, % weight gain and production of supermale tilapia were significantly ($p<0.05$) higher than those of monosex tilapia. But SGR (% day), FCR and survival rate of supermale tilapia were not significantly ($p>0.05$) varied. However, the result of the present study showed that the best weight gain of 124.85 g was observed in ST_1 after 90 days culture period. Average weight gain (g) were 1.39, 1.16, 1.14 and 1.05, SGR (per day) were 2.17, 2.09, 2.08 and 2.04%, FCR were 2.98, 2.65, 2.84 and 2.57, survival rate were 96, 94.50, 95 and 91% and fish production were 5053.92, 8926.10, 4108.07 and 7821.41 kg/ha in ST_1, ST_2, MT_1 and MT_2 respectively. The present research findings suggested that supermale tilapia has significantly high growth potential compare to monosex tilapia under mini ponds culture condition.

INTRUDUCTION

Tilapia is known to be an important of subsistence fisheries for thousands of years but have gained prominence in recent years. Tilapia, that is native to Africa and Middle East, has emerged from mere obscurity to one of the most productive and internationally traded food fish in the world. The introduction of the tilapia in Bangladesh from Thailand was first initiated in 1954 with *Tilapia mossambica* (Ahmed, 1956) and later in 1974, high yielding species of tilapia (*Oreochromis niloticus*) was introduced by UNICEF (Rahman, 1985) with a hope that it would make a significant contribution to fish production. *Orechromis niloticus* has for many decades, been responsible for the significant increase in global tilapia production from freshwater aquaculture and accounted for about 83% of total tilapias produced worldwide (FAO, 2002). Monosex tilapia (*Oreochromis nloticus*) newly introduced as exotic species in aquaculture system of Bangladesh (FAO, 1999).

Many potential rural fish farmers and pond owners of Bangladesh are poor and they do not have the capability to invest much money for purchasing fish seed, fertilizer and feed. As a result their ponds remained derelict. The Department of Fisheries (DoF, 1993) estimated about 17% of derelict ponds or ditches in Bangladesh, which are lying fallow, expect some used for catching wild fishes only. These derelict ditches retain water for 4-6 months and can be utilized properly by culturing short cycle species like monosex tilapia (*Orechromis niloticus*). In Bangladesh aquaculture is the most promising option for increasing fish production. Monosex tilapia is the best candidate to overcome this situation due to its desirable characteristics such as males are used for monosex culture grows faster than females (Popma and Lovshin, 1996). Male monosex culture permits the use of longer culture periods, higher stocking rates and fingerlings of any age. Monosex tilapia has good resistance to poor water quality, disease and tolerance to a wide range of environmental conditions.

Monosex tilapia is a fast growing popular cultivable fish (Chowdhury *et al.*, 1991). In Bangladesh, commercial farming of tilapia has been found to develop rapidly since the introduction of Genetically Improved Farmed Tilapia (GIFT) from the Philippines in 1994 (Alam and Kawsar, 1998). The success of using the GIFT strain of tilapia for commercial farming is due to its ability to produce millions of monosex male fry in hatcheries and this practice has been found to considerably eliminate the problems related to the production of mixed sex tilapia showing slow growth as well as the production of small-sized individuals in a given culture facility (Mair and Little, 1991). Recently farmers of Mymensingh region introduced supermale (YY male) tilapia through genetic manipulation (Haque, 2012 personal communication). It is thus important to investigate the culture potential of this tilapia and compare with the growth of monosex tilapia. Therefore, the present experiment has been designed to study culture potential and growth variation of supermale tilapia with those of monosex tilapia and culture feasibility of supermale tilapia in earthen mini ponds.

MATERIALS AND METHODS

The experiment was conducted in eight experimental pond each of 0.65 dec which were located in the northern side of the Faculty of Fisheries, Bangladesh Agricultural University, Mymensingh. The study period was carried out for 90 days from 03 May to 31 July, 2012. The water depth was maintained at a level of 1.0 to 1.3 m. The ponds were equal in size and similar in shape, depth, basin configuration and pattern type including water supply facilities. Aquatic weeds, undesirable fishes, insects and other aquatic organisms were removed manually and the grasses on the pond dykes were also pruned manually into very small size. Lime was applied at a rate of 0.5 kg/dec. No fertilizer was used during pond preparation.

Two treatments were considered for supermale tilapia and two treatments for monosex tilapia. In each treatment two replications were considered. Monosex tilapia fry were collected from Sarnalata Agro Fisheries Ltd., Radhakanai, Fulbaria, Mymensingh and super male tilapia fry were collected from the Brahmaputra Hatchery, Shamvuganj, Mymensingh. Fry were transported by plastic drums having oxygen facilities and transferred to ponds. During stocking sufficient care was taken to reduce stress. Commercial pellet feed named "Quality Fish Feed" were selected for the study. At the beginning of the experiment feed was supplied at a rate of 30% (1st month) of their body weight, 15% (2nd month), 5% up to harvesting time. Half of the feed was supplied at 9:00 AM and remaining was supplied at about 5:00 PM.

The feed was supplied by spreading method. The experimental ponds were monitored everyday during feeding to observe the behavior of fishes. All the ponds were kept clean to provide hygienic condition. Water quality parameters such as temperature (ºC), dissolved oxygen (mg/l), pH and ammonia (mg/l) were recorded fortnightly. Parameters such as weight gain (g), average daily weight gain (g), percent weight gain, specific growth rate (SGR), food conversion ratio (FCR), survival rate (%) and production (kg/ha/yr) were calculated to evaluate the growth performances of fish. Fish sampling was done at fifteen days interval in the morning at around 7:30 AM to 8:30 AM. During each sampling, fish were caught by cast net and weight was taken by precision weighing balance. Data were kept for analysis of different parameters.

RESULTS AND DISCUSSION

The mean initial weight of supermale tilapia and monosex tilapia in both the treatments was 1.4 g. Mean weight gains of supermale at the end of the experiment were 124.85 g and 104.35 g and monosex tilapia were 102.35 and 94.45 g in T_1 and T_2 respectively (Table 1 and Fig. 1).

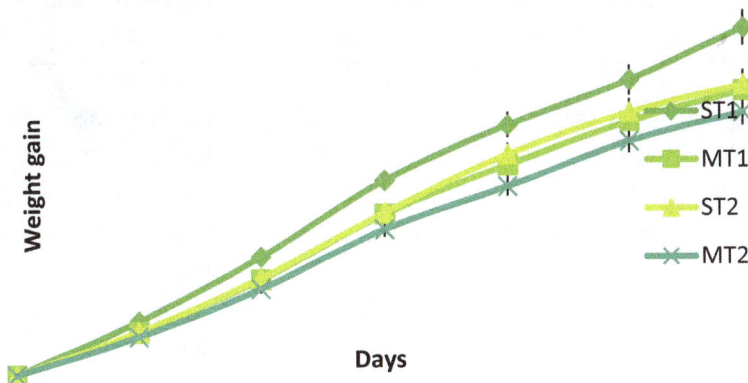

Figure 1. Mean weight gain of both tilapia in both treatments during the experimental Period

Average daily weight gain of supermale and monosex tilapia at the end of the experiment was followed by 1.39, 1.16, 1.14 and 1.05 g in ST_1, ST_2, MT_1 and MT_2, respectively. Mean weight gain of monosex tilapia for 90 days experiment were 102.359 and 94.45 g in T_1 and T_2, respectively. The percent weight gain of supermale tilapia till the end of the experiment were 8917.86 and 7453.57 in T_1 and and T_2, repetively (Table 1). In monosex tilapia the percent weight gain was 7310.71 and 6746.43 in T_1 and T_2, respectively. During the investigation of 90 days specific growth rate (SGR) in T_1 of supermale and monosex were 2.17% and 2.08% (Table 1). In T_2 of supermale and monosex tilapia the values were 2.09% and 2.04% respectively (Table 1). The mean values of FCR for supermale in T_1 and T_2 were 2.98 and 2.65 respectively (Table 1). In case of monosex tilapia, the mean values of FCR were 2.84 and 2.57 respectively (Table 1). The survivals of fish at the end of the experiment were 96±1.00, 94.5±0.50, 95±1.00 and 91±1.00 in ST_1, ST_2, MT_1 and MT_2 respectively. Higher survival rate were obtained in ST_1 (96%) and lower was in MT_2 (91%) (Table 1). Production of supermale tilapia in T_1 and T_2 was 5063.916 kg/ha and 8926.099 kg/ha and production of monosex tilapia in T_1 and T_2 were 4108.072 kg/ha and 7821.405kg/ha respectively (Table 1).

The range of water temperature were 28.25 to 30°C in ST_1, 28.63 to 31.25°C in ST_2, 28.13 to 31.75°C in MT_1 and 28.33 to 30.25°C in MT_2. Range of dissolved oxygen values were 6.5 to 8 mg/l, 7 to 8 mg/l, 6.5 to 7.65 mg/l and 6.5 to 8 mg/l in ST_1, ST_2, MT_1, and MT_2. The range of pH values were recorded from 7.25 to 7.75 in ST_1, 7.25 to 7.38 in ST_2, 7.25 to 7.88 in MT_1 and 7.08 to 7.88. The ammonia content of the experiment was varied from 0.15 to 0.25 mg/l in ST_1, 0.18 to 0.30 mg/l in ST_2, 0.18 to 0.33 mg/l in MT_1 and 0.20 to 0.25 mg/l in MT_2.

Figure 2. Supermale Tilapia

Figure 3. Monosex Tilapia

Table 1. Growth parameters of supermale and monosex tilapia in ST_1 and MT_1 during the study period

Growth parameters	Supermale Tilapia ST$_1$	Monosex Tilapia MT$_1$	Supermale Tilapia ST$_2$	Monosex Tilapia MT$_2$
Mean initial weight (g)	1.40 ± 0.09^a	1.40 ± 0.09^a	1.40 ± 0.09^a	1.40 ± 0.09^a
Mean final weight (g)	126.25 ± 4.41^a	103.75 ± 3.42^b	105.75 ± 4.06^b	95.85 ± 4.35^c
Mean weight gain (g)	124.85 ± 12.29^a	102.35 ± 4.76^b	104.35 ± 6.77^b	94.45 ± 6.37^c
Av. daily weight gain (g)	1.39 ± 0.14^a	1.14 ± 0.05^b	1.16 ± 0.08^b	1.05 ± 0.07^c
% weight gain	8917.86 ± 877.61^a	7310.71 ± 340.30^b	7453.57 ± 483.58^b	6746.43 ± 454.93^c
SGR (%/day)	2.17 ± 0.05^a	2.08 ± 0.02^a	2.09 ± 0.03^a	2.04 ± 0.04^a
FCR	2.98 ± 0.57^a	2.84 ± 0.83^a	2.65 ± 0.19^a	2.57 ± 0.83^a
Survival rate	96.00 ± 1.00^a	95.00 ± 1.00^a	94.50 ± 0.50^a	91.00 ± 1.00^a
Production	5063.916 kg/ha^a	4108.072 kg/ha^b	8926.099 kg/ha^c	7821.405 kg/ha^d

*Superscripts in each row with different letter significantly (P<0.05) different

In the present study the highest of weight gain significantly (p<0.05) higher was found in T_1 (124.85 g) of supermale tilapia whereas, the lowest weight gain was found from the T_2 (94.45 g) of monosex tilapia. Also from T_2 of supermale, final weight was found significantly (p<0.05) higher (104.35 g) when compared with T_1 (102.35 g) and T_2 (94.45 g) of monosex tilapia. The results indicated that the growth rate of supermale tilapia was higher than monosex tilapia. Supermale tilapia had been reported to grow faster than the mixed-sex tilapia or the monosex tilapia obtained from the conventional hormone induction method (Mair and Little, 1991; Rahman and Sarder 2010). Ahmed *et al.* (2013) obtained a weight gain of 123.48 g and 111.82 g from two different treatment of monosex tilapia for a period of 70 days which were higher than the values obtained from the present investigation. The results of the present experiment showed that the growth rate of tilapia in both the variety were higher in lower stocking densities. Begum (2009) obtained 47.03 g and 39.93 g of monosex tilapia at stocking densities of 200 fish/dec and 300 fish/dec respectively which were much lower than the value obtained in the present study. It was observed in the present experiment that the highest mean weight gain (124.85 g) of fish was in ST_1 stocked at lower densities, although, same feed and feeding rate were applied in all the treatments. The mean lowest weight gain (94.45 g) obtained in the present experiment under the highest stoking rate of 200/dec in MT_2. Kohinoor *et al.* (1998) found the highest growth of tilapia stocked at the rate of 80 fish/dec which were much lower than the value obtained in the present investigation. At the end of the experiment the highest average daily weight gain was in ST_1 (1.39 g/day) and the lowest average daily weight gain was in MT_2 (1.05 g/day). According to Rasel (2012) the recorded mean average daily weight gain was 0.0049 g/day, 0.0062 g/day and 0.0073 g/day during the rearing of fry of monosex tilapia in three different treatments for 100 days which were lower than present findings. From the research findings of Das (2007) it was observed that the highest average daily weight gain was 1.94 g/day for *Oreochromis niloticus* fed on formulated diet which were higher than the values obtained in the present study.

From the investigation the highest mean percent weight gain (8917.86 g) was found from the ST_1 compared to ST_2 (7453.57 g), MT_1 (7310.71 g) and MT_2 (6746.43 g). This might be due to less competition for feed in lower stocking density and also for higher growth rate of supermale tilapia. From the research findings of Ahmed *et al.* (2013) it could be mentioned that percent weight gain for monosex tilapia were 123348.44 g and 11181.78 g in T_1 and T_2 respectively which were higher than present findings. Begum (2009) found the highest mean percent weight gain for monosex tilapia was 9406 g and 7986 g for four months culture period in T_1 and T_2 respectively which has similarity with the present findings.

The result of the present experiment revealed that values of SGR of supermale were 2.17% in T_1 and 2.09% in T_2. In monosex tilapia SGR values were 2.08% and 2.04% in T_1 and T_2 respectively. Higher values of SGR were obtained from ST_1 and MT_1 which had lower stocking densities. Islam (2007) and Alam (2009) obtained the highest values of SGR at the lowest stocking densities which coincide with the present findings. According to Mamun *et al.* (2010), Genetically Male Tilapia and Sex Reversed Tilapia were grown in six earthen ponds and found SGR values as 0.997% and 0.988% respectively, which were much lower than the value obtained in the present investigation. Hossain *et al.* (2004) observed SGR values of tilapia were ranged from 2.04 to 2.30 fed on formulated diet that has similarity with the findings of present study. In the present investigation FCR were varied from 2.98 to 2.57.

FCR values for supermale tilapia in T_1 and T_2 were 2.98 and 2.65 respectively, whereas, for monosex tilapia in T_1 and T_2 were 2.57 and 2.84 respectively. From the research finding of Ahmed *et al.* (2013) it was observed that FCR values for monosex tilapia fed on homemade feed were 1.51 and 1.40 in T_1 and T_2 respectively. Hossain *et al.* (2004) investigated FCR of gift strain of tilapia fed on formulated diet (30.09% protein) was 1.71 and 1.77 which was lower than present results. The fishes might have properly utilized most of the formulated feed and the utilized feed help in production of supermale and monosex tilapia in the present study.

In the present experiment the highest survival rate was recorded in T_1 of supermale tilapia and the lowest survival rate in T_2 of monosex tilapia. Kohinoor *et al.* (2007) observed survival rates of monosex tilapia were varied from 79% to 92%. According to Ahmed *et al.* (2013) survival rate of monosex tilapia were 75.55% and 90.37% in T_1 and T_2 respectively during the harvesting time which has similarity with the survival rate of present experiment. Survival rate was found to be negatively influenced by different stocking densities such as the lowest stocking density showed the highest survival rate which might be due to high competition of food and space among the fishes.

The productions of supermale and monosex tilapia were 5053.92 kg, 8926.1 kg, 4108.07 kg and 7821.41 kg in ST_1, ST_2, MT_1 and MT_2 respectively. Although mean weight gain was found higher in ST_1 but total production was higher in ST_2 which might be due to higher number of fishes. The present result supports the findings of Roy (2002) who achieved the best production from higher stocking densities in comparison with that achieved with the lower ones. Ahmed *et al.* (2013) mentioned that average yield of monosex tilapia were 19076 kg/ha and 16312.11 kg/ha in two different treatments with same stocking densities fed with homemade feed and formulated feed in earthen mini pond. Begum (2009) observed the highest production was 9.63 kg/dec/120 days whereas Das (2007) found total production was 34.04 kg/dec/90 days for monosex tilapia.

A simple economic analysis of the growth performance of fish showed that the highest net profit (Tk/ha/year) of Tk 9,56,685 was obtained with ST_2. The highest net profit in ST_2 was due to high growth rate and density tolerance of supermale tilapia compared to monosex tilapia. Das (2007) mentioned that the net profit of *Oreochomis niloticus* was 3,87,716 Tk/ha/year using formulated diet which were much lower than the value obtained in the present investigation. According to Chakma (2011) net benefit were 2,07,328.53 Tk/ha/70 days and 77,917 Tk/ha/70 days.

Culture of monosex tilapia bears high production potential in Bangladesh. A number of hatcheries are producing monosex male fry through androgenic hormone feeding, but a considerable percentage of female fry in each batch has been reported. Supermale tilapia can eliminates the customers concern on residual hormonal health hazard. From the experiment it was found that supermale had high growth rate and density tolerance compare to monosex tilapia. Supermale tilapia has nice reddish color and shape is slightly round which attract the customer's attention. During the economic analysis it was found that the net profit from the present experiment might be due to higher mean increased weight of supermale tilapia. So it can be brought to a conclusion that the supermale tilapia has high growth potential in comparison with monosex tilapia.

Proper training on culture of this new variety on mini ponds could help in poverty alleviation at rural farmer's level. Income of the rural poor farmers could be increased through derelict household mini ponds brought under cultivation. All family members especially women can take participation in supermale tilapia culture and could contribute to family income.

REFERENCES

1. Ahmed GU, N Sultana, M Shamsuddin and MB Hossain, 2013. Growth and Production Performance of Monosex Tilapia (*Oreochromis niloticus*) Fed with Homemade Feed in Earthen Mini Ponds. Pakistan Journal of Biological Sciences, 1-5.

2. Ahmed M, 1956. Transplantation of fish food to East Pakistan. Pakistan Journal of Science, 8: 167-170.

3. Alam MN, 2009. Effects of stocking density on the growth and survival of monosex male tilapia (*Oreochromis niloticus*) fry (GIFT Strain) in hapa, MS Thesis, Department of Aquaculture, Bangladesh Agricultural University, Mymensingh. pp. 40.

4. Alam SM and MA Kawsar, 1998. Effect of estrogens on growth and sex-ratio in the genetically imporved farmed tilapia, *Oreochromis niloticus* (L.). Bangladesh Journal of Zoology, 26: 37-43.

5. Begum M, 2009. Effects of stocking density on the growth and production performance of monosex tilapia, MS Thesis, Department of Aquaculture, Bangladesh Agricultural University, Mymensingh. pp. 31.

6. Chakma A, 2011. Growth performance of Thai pangus (*Pangasianodon hypophthalmus*) using prepared and commercial feed, MS, Thesis, Department of Aquaculture, Bangladesh Agricultural University, Mymensingh. pp. 49.

7. Chowdhury MBR, MM Muniruzzaman and N Uddin, 1991. Studies on the intestinal bacteria; flora of tilapia, *Oreochromis niloticus*. Bangladesh Journal of Aquaculture, 11: 23-25.

8. Das R, 2007. Effect of stocking density on the growth, survival and production of monosex male tilapia (*Oreochromis niloticus)* fed on formulated diet, MS Thesis, Department of Aquaculture, Bangladesh Agricultural University, Mymensingh. pp. 77.

9. DoF, 1993. Fish catch statistics of Bangladesh. 1992-93. Department of Fisheries, Dhaka, Bangladesh. (BGP-94/95-5177B-1000-24-7-95).

10. FAO, 1999. Fisheries Statistics, Capture Production. Food and Agriculture Organization of the United Nations (FAO), Rome. pp. 703.

11. FAO, 2002. Fishery Statistics. Aquaculture production, 90(2).

12. Hossain MA, R Roy, SM Rahmatullah and AHM Kohinoor, 2004. Effect of stocking density on the growth and survival of GIFT tilapia, (*O. niloticus*) fed on formulated diet. Journal of Agriculture and Rural Development, 2: 127-133.

13. Haque MN, 2012. Supermale (YY) tilapia. Owner of Brahmaputra Hatchery, Shamvuganj, Mymensingh.

14. Islam MS, 2007. Effects of stocking density on the growth and production performance of Tilapia (*Oreochromis niloticus*) in ponds, MS Thesis, Department of Aquaculture, Bangladesh Agricultural University, Mymensingh. pp. 54.

15. Kohinoor AHM, AKMS Islam, DA Jahan, M Zakir and MG Hussain, 2007. Monoculture of climbing perch, Thai koi, (*Anabas testudineus*) (Bloch) under different stocking densities at on farm. Bangladesh Journal of Fisheries Research, 11: 173-180

16. Kohinoor AHM, PC Modak and MG Hussain, 1998. Growth and production performance of red tilapia and Nile tilapia under low input culture system. Bangladesh Journal of Fisheries Research, 3: 11-17.

17. Mair GC and DC Little 1991.Population control in farmed tilapia, Naga, ICLARM Quarterly. pp. 8-13.

18. Mamun AA, MRI Sarder and MM Rahman, 2010. Growth performance of genetically male (GMT) and hormone induce sex reversed male tilapia (*Oreochromis niloticus*, L.) in earthen pond aquaculture system. Bangladesh Journal of Zoology, 38: 163-169.

19. Popma TJ and LL Lovshin, 1996. Worldwide prospects for commercial production. Research and Development Series No. 41. Department of Fisheries and Allied Aquaculture, Auburn University, Alabama, USA. pp. 23.

20. Rasel M, 2012. Studies on brood rearing fry production and nursing of monosex tilapia, MS Thesis, Department of Aquaculture, Bangladesh Agricultural University, Mymensingh. pp. 50.

21. Rahman MM and MRI Sardar, 2010. Production of Hormone-induced Supermale of Genetically Improved Farmed Tilapia (GIFT-YY) in Bangladesh. Asian Fisheries Science, 23: 136-144.

22. Rahman AKA, 1985. Introduction of exotic fishes in Bangladesh. Paper presented in the seminar on culture of exotic fish in Bangladesh. October 15, 1989. Bangladesh Zoological Society. Dhaka University, Dhaka, 14.

23. Roy R, 2002. Effect of stocking density on the growth and survival of GIFT tilapia fed on formulated diet. MS Thesis, Department of Aquaculture, Bangladesh Agricultural University, Mymensingh. pp. 63.

4

COMPARATIVE SHELF LIFE STUDY OF WHOLE FISH AND FILLETS OF CULTURED STRIPED CATFISH (*Pangasianodon hypophthalmus*) DURING ICE STORAGE CONDITION

Salma Noor-E-Islami[1], Md. Faisal[2*], Mousumi Akter[3], Md. Shaheed Reza[4] and Md. Kamal[4]

[1]Department of Fisheries (DoF), Government of the People's of Republic Bangladesh; [2]Department of Fishing and Post Harvest Technology, Faculty of Fisheries, Chittagong Veterinary and Animal Sciences University, Chittagong-4225; [3]Department of Fisheries Technology, Faculty of Fisheries, SFM Fisheries College, Melandah, Jamalpur-2010; [4]Department of Fisheries Technology, Faculty of Fisheries, Bangladesh Agricultural University, Mymensingh-2202, Bangladesh

*Corresponding author: Md. Faisal, E-mail: faisalmohammad10@gmail.com

---

ARTICLE INFO

ABSTRACT

Key words

Striped catfish
Shelf life
Fish fillet
Ice storage

A comparative study on the shelf life of whole fish and fillet of cultured striped catfish (*Pangasianodon hypophthalmus*) under ice storage condition was carried out by determining organoleptic and biochemical aspects in the month of September and October. It was found that shelf life of whole fish was 21 days where as fillet had slightly lower shelf life of 18 days. Value of pH changed from 7.07 to 6.97 after 24 days for whole fish and 6.88 for fillet after 21 days at the termination of experiment, pH remained to 6.5 for fillet after 18 days of ice storage. TVB-N value increased from an initial value of 1.68 mg/100g to 28.83 mg/100g in 21 days in whole fish and finally to rejection value of 35.89 mg/100g at the end of 24 days storage period. As for fillets TVB-N value was at acceptable level of 24.79 mg/100g until 18 days of storage at same condition. Myofibriller protein solubility gradually decreased with storage period for both whole fish and fillets. The results of this study indicated that in regard to organoleptic and biochemical aspects the shelf life of *Thai-Pangas* in ice stored condition is 21 days and 18 days for the whole fish and fillets, respectively.

INTRUDUCTION

There has been a rising demand of chilled-stored striped catfish (*Pangasianodon hypophthalmus*, locally called *Thai-Pangas*) in Bangladesh and other parts of the world over the last decade. Its consumption has grown rapidly in Russia, Spain, France and other European countries and the United States. In Europe, the presence of *Pangasianodon sp.* is overwhelming and there is practically no fish shop without *pangas* fillets. *Thai-Pangas* catfish is a low-priced fish in the common fish market. It is cultured due to its faster growth, ease of reproduction, adaptability to intensive culture, acceptability of low input sustainable feeds, resistance to impaired water quality, good taste, fewer bone content and widespread consumer acceptance (Ali et al., 2013; De Silva, 2011;).

After death *Thai-Pangas* is highly susceptible to spoilage from post-mortem autolysis and microbial growth. This accounts for use of appropriate storage method for preservation of its quality. There are several methods to preserve fresh fish; however, icing is a common method for short-term preservation. During icing or chilling storage of fish, chemical changes are known to take place. There are a number of factors that influence the quality of fish. Of which the most important one is the post-mortem changes that take place soon after death due to enzymatic action. The state of rigor in association with other biochemical changes influences the meat quality in fish and higher animals.

As such, there has been little or no study conducted in Bangladesh on the shelf life of this commercially important fish species. Although Hossain et al. (2005) studied some physiochemical changes in whole *Thai-Pangas* muscle during ice storage condition, there are no reports on different changes in its fillets under different storage conditions. This study was, therefore, carried out to investigate the shelf life of whole fish and fillets of *Thai-Pangas* by determining organoleptic and biochemical changes under ice storage condition.

MATERIALS AND METHODS

Storage condition and fish sampling

Thirty (30) live *Thai-Pangas* with average length of 43 cm and body weight of 1.25 kg were collected from 2 fish farms of Trishal and 1 fish farm of Bhaluka, Mymensingh and transported to the Laboratory of Fish Processing, Department of Fisheries Technology, Bangladesh Agricultural University, Mymensingh. Fish were killed with cranial spiking, washed with tap water, placed in plastic bags and kept in ice until subsequent experiments.

Organoleptic assessment

Sensory evaluation was carried out according to the guidelines described by EC freshness grade for ice stored fish (Howgate and Whittle, 1992).

Analytical methods

Measurement of muscle pH

During the determination time of rigor index 2 g of the ordinary muscle was dissected from the dorsal part of the sample and homogenized with 10 ml of distilled water in a blender. After adjusting the temperature, the electrode of the pH meter was put on the tube and the pH was measured by the pH meter (Model 250 pH/ISE).

TVB-N Value

TVB-N values were determined as described by Antonacopoulos and Vyncke (1989).

Changes in the protein solubility

Myofibrils are a part of muscle protein. Myofibrillar protein solubility indicates the denaturation rate of protein.

Preparation of myofibrils

Myofibrils were prepared from ordinary muscles immediately after excision according to Perry and Grey (1956) with slight modification. The muscle was chopped by a meat grinder and chilled minced muscle (50 g) was homogenized for 1 min in 5 volumes of 39 mM borate buffer (pH 7.1) containing 25 mM KCl and 0.1 mM DTT. The homogenate was centrifuged for 15 min at 600×g. The precipitate obtained was again homogenized and centrifuged for 15 min at 600×g. The light colored upper layer of the precipitate consisting mainly of myofibril was recovered with small volume of 39 mM borate buffer (pH 7.1) containing 0.1 M KCl and 0.1 mM DTT. The suspension was centrifuged for 15 min to remove the supernatant. Myofibrils were diluted with 4 volume volumes of 39 mM borate buffer (pH 7.1) containing 0.1 M KCl and 0.1 mM DTT and coarse materials were removed by centrifugation at 400×g. The suspension was centrifuged for 15 min at 600×g to precipitate myofibril. After the precipitate was washed three times in the same way, myofibril were suspended with desired volume of 39 mM borate buffer (pH 7.1) containing 0.1 M KCl to make a concentration of 10-15 mg/ml.

Myofibrillar protein solubility

Two ml of myofibrillar suspensions (5mg/ml) were homogenized with 2 ml of 1M KCl plus 100mM phosphate buffer (pH 7.0) using a homogenizer. The homogenate was allowed to stand at refrigerated temperature (4°C) overnight. The suspension was centrifuged for 30min at 900×g in cold condition and the supernatant was collected. Protein concentration was determined by Biuret method (Gornall et al., 1949).

Statistical analysis

Data from different biochemical measurements were subjected to t-test ($p<0.05$). Statistical software package SPSS 10.1 (SPSS Inc, Chicago, IL, USA) was used to explore the statistical significance of the results obtained.

RESULTS AND DISCUSSION

Changes in organoleptic quality under ice storage condition

The quality of fishes was graded by using the score from 1-5. The score points less than 2 were considered as excellent. The points from 2 to less than 5 were judged as good or acceptable conditions, while 5 and above considered as bad or rejected. The changes in quality of chilled fish during storage were assessed by daily organoleptic examination. Table 1 showed the changes in organoleptic qualities of whole fish and fillets of *Thai-Pangas* during ice storage in an insulated box. On the basis of the scores the fishes were found in acceptable conditions for 21 days of ice storage before becoming inedible. Hossain et al. (2005) reported that shelf life of *Thai-Pangas* muscle was 20 days under storage at insulated box in ice which is also more or less similar with the findings of present study. Reza et al. (2009) reported that the quality changes of tropical fishes in ice takes place in 4 phases corresponding to periods of 0 to 3, 4 to 6, 7 to 9 and 10 to 13 days. Similarly in the present study, it was observed that very little change occurred in whole fish and fillets during 0 to 3 days and 3 to 6 days without loss of natural flavor and odor in *Thai-Pangas*. During this period whole fish and fillets had the characteristics of excellent quality. During the periods of 6 to 9 days, 9 to 12 days and 12 to 15 days, In case of whole fish there were little deterioration apart from slight loss of natural flavor and odour, and partial loss of bloom. At this stage there was little loss of the characteristic odour and the flesh was neutral but had no off-flavour. Fillets showed the early spoilage with sour flavour during 12 to 15 days of storage whereas in case of whole fish, the early signs of spoilage was observed during 15-18 days of storage. At the beginning of this phase the flavor in whole fish and fillet was slightly sour, sickly sweet, fruity of like dried fish, which was judged as acceptable quality. The taste began to stale, as well as the appearance and texture began to show the obvious signs of spoilage for fillet under acceptable condition during the period of 15-18 days. On the other hand the taste, texture and appearance of whole fish showed the evidence of primary spoilage including the gills and the belly cavity had an unpleasant smell in the limit of acceptance during 18 to 21 days. The fillet became putrid by all of the characteristics and hence rejected during day 18 to 21 days, whereas the whole fish became rejected during 21 to 24 days for the same reason.

In the present study, iced stored *Thai-Pangas* was found to be acceptable up to 21 days and the fillets were acceptable up to 18 days indicating significantly lower ($p<0.05$) shelf life of fillets than whole fish from organoleptic point of view. While fish from the tropics had significantly higher storage life under ice compared to those from the temperate region, higher shelf life of fish fillet was obtained for *Thai-Pangas* when compared to fillet of cold water fishes (Onibi et al., 1996; Gatta et al., 2000; Pirini et al., 2000; Ruff et al., 2002). Although the present study did not consider the development of color and related color index with storage period, its use could have reduced the shelf life of *Thai-Pangas* fillets to further 1 to 2 days.

Table 1. Changes in organoleptic qualities of whole fish and fillets of *Thai-Pangas* during ice storage condition

Days of storage	Mean defect points		Grade		Overall qualities	
	Whole fish	Fillet	Whole fish	Fillet	Whole fish	Fillet
0	1.25	1.25	A	A	Excellent	Excellent
3	1.57	1.57	A	A	Excellent	Excellent
6	2.00	2.00	A	A	Excellent	Excellent
9	2.20	2.20	B	B	Acceptable	Acceptable
12	2.9	2.9	B	B	Acceptable	Acceptable
15	3.45	3.97	B	B	Acceptable	Acceptable
18	3.97	4.6	B	B	Acceptable	Acceptable
21	4.6	5.0	B	C	Acceptable	Rejected
24	5.0	-	C	-	Rejected	-

Changes in pH value

Changes in muscle pH of ice stored whole fish and fillets are shown in (Fig.1). The pH of fish muscle immediately after death was around 7.07. After 6 days of storage in ice, the pH decreased gradually to 6.11 in case of whole fish and 6.15 in case of fillets. After 9 days of ice storage pH increased gradually due to formation of basic compounds and finally reached up to 6.97 after 24 days and 6.88 after 21 days for fillets when the whole fish and fillet were organoleptically unacceptable. The correlation between pH and organoleptic scores in this study suggests that pH can be used as a reliable index of quality. Low muscle pH of the post-mortem fish muscle is associated with the quality changes in fish (Konagaya and Konagaya, 1979; Kramer and Peters, 1981; Reza et al., 2009). Muscle pH of live sardine has been reported as 7.2 (Pacheco-Aguiler et al., 2000) but following the death, muscle pH decreased to 6.8 after 2 h, 6.2 after 8 h and 5.8 after 24 h, respectively (Watabe et al., 1991). The pH values of sardine muscle stored in ice was 5.8, 6.36 and 6.57 at day 0, 9 and 18, respectively (E1 Marrakch et al., 1990).

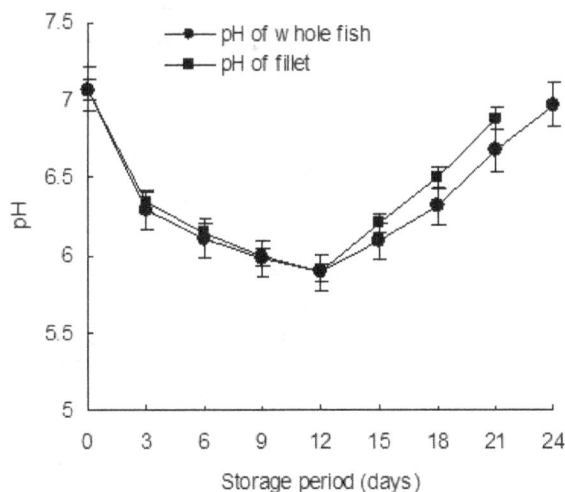

Figure 1. Changes in muscle whole fish and fillets of *Thai-Pangas* during ice storage

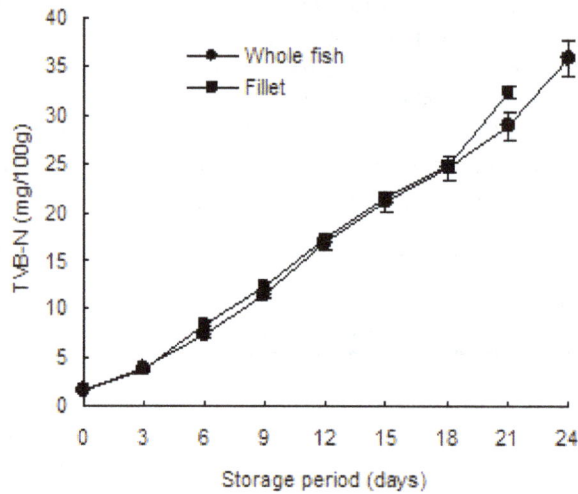

Figure 2. Comparison of changes in TVB-N of whole fish and fillets of *Thai-Pangas* during ice storage

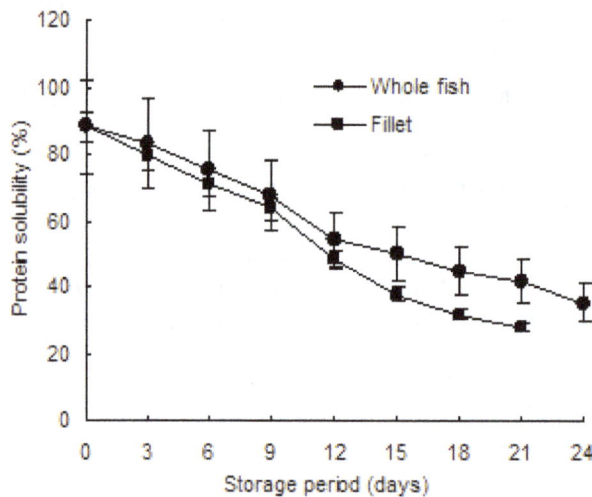

Figure 3. Comparison of changes in muscle protein solubility of whole fish and fillets of *Thai-Pangas* during ice storage

Changes in TVB-N value

The amount of TVB-N in fish increases as spoilage progresses. Increase in TVB-N with the lapse of storage may be attributed to bacterial spoilage. The results of the TVB-N (mg/100g) of experimental whole fish and fillets have been presented in Fig. 2. The initial TVB-N values were 1.68 mg/100g in *Thai-Pangas*, which gradually increased with lapse of storage period. At the end of 21 days of ice storage TVB-N value of whole fish increased to 28.83 mg/100g whereas the TVB-N value of fillets increased to 24.79 mg/100g at the end of 18 days of storage. These values are within the range of recommended values of 25 to 30 mg/100g for fresh fish (Connell, 1980). However, at the end of 24 days of ice storage the TVB-N value of whole fish was 35.89 mg/100g and at the end of 21 days of ice storage the TVB-N Value of fillet was jumped to 32.45 mg/100g due to combined action of enzymes and microbial activity and thereby exceeded the acceptable limit. It has been reported that TVB-N value mainly increased in the fish flesh during the later phase of spoilage when the bacterial population increased (Mazorra-Manzano et al., 2000). TVB-N values showed a slight increase for whole ungutted sea bass during storage, reaching a value of 26.77 mg N per 100 g muscle (day 13), whereas for fish fillet corresponding value of 26.88 mg N per 100 g muscle was recorded (day 9) (Taliadourou et al., 2003). Thus TVB-N value is low during the edible storage period and increased amount of TVB-N are found near the rejection state of fish.

Changes in protein solubility

Changes in protein solubility of whole fish and fillets of *Thai-Pangas* during ice storage have been presented in Fig. 3. Solubility of myofibrillar protein immediately after death was 88.21%, which decreased gradually to 35.27% in case of whole fish at the end of 24 days of storage whereas at the end of 21 days of storage the solubility of myofibrillar proteins from fillet was 28.12%. The solubility of carp myofibrils decreased from 95% to 20% during ice storage within 2-3 weeks (Seki et al. 1979). Similar findings were also reported for for mrigal (*Cirrhina mrigala*) (Hossain, 1995) and ruhu (*Labeo rohita*) (Faruk, 1995). According to Seki et al. (1980) the fall in solubility during ice storage was due to lowering of pH. The loss in myofibril solubility of milkfish during storage was due to aggregate formation by disulfide, hydrogen and hydrophobic bonds (Jiang et al., 1988). The solubility in muscle protein of croaker, lizardfish, threadfin bream and big-eye snapper decreased continuously during prolonged storage. The results obtained from the present study indicated that the decrease in solubility of pangas muscle might be due to aggregation as well as denaturation of muscle protein during ice storage.

CONCLUSION

Cultured *Thai-Pangas* fillet samples showed a significantly shorter shelf life (18 days) compared to its counterpart whole fish (21 days) samples in ice storage condition. Fillets were found to be more prone to oxidation than the counterpart whole fish species, because of a greater expouser of the fish muscle in fillets to oxygen. If longer shelf-life times are commercially required for fillets, protective treatments such as vacuum packaging, modified atmosphere packaging and natural antioxidant application are recommended. Improvements in the shelf life of a product can have important economic impact by reducing losses attributed to spoilage and by allowing the products to reach distant new markets and in this fashion very helpful to prepare value added products.

REFERENCES

1. Ali H, MM Haque and B Belton, 2013. Striped catfish (*Pangasianodon hypophthalmus*, Sauvage, 1878) aquaculture in Bangladesh: an overview. Aquaculture Research, 44: 950–965.
2. Antonacopoulos N and W Vyncke, 1989. Determination of volatile basic nitrogen in fish: a third collaborative study by the West European Fish Technologists' Association (WEFTA). Zeitschrift Fur Lebensmittel-Untersuchung Und-Forschung, 189: 309-316.
3. Connell JJ, 1980. Quality deterioration and extrinsic quality defects in raw material. In: control of fish quality, 2nd, ed. Fishing News Books Ltd. Surrey, England. pp. 31-35.
4. De Silva SS and NT Phuong, 2011. Striped catfish farming in the Mekong Delta, Vietnam: a tumultuous path to a global success. Reviews in Aquaculture, 3: 45–73.
5. El Marrakch A, M Bennour, N Bouchriti, A Hamama and H Tagafatit, 1990. Sensory, chemical and microbiological assessments of Moroccan sardine (*Sardina pillchardus*) stored in ice. Journal of Food Protect, 53: 600-605.
6. Faruk MAR, 1995. Studies on the post-mortem changes in rohu fish (*Labeo rohita*). M. Sc. Thesis, Department of Fisheries Technology. Bangladesh Agricultural University, Mymensingh.
7. Gatta PP, Pirini M, Testi S, Vignola G and Monetti PG, 2000. The influence of different levels of dietary vitamin E on sea bass *Dicentrarchus labrax* flesh quality. Aquaculture Nutrition 6: 47–52.
8. Gornall AG, CJ Bardawill and MM David, 1949. Denaturation of serum proteins by means of the biuret reaction. Journal of Biological Chemistry, 751-766.
9. Hossain MI, 1995. Studies on the post-mortem changes in mrigal (*Cirrhina mrigala*). M.Sc. Thesis. Department of Fisheries Technology. Bangladesh Agricultural University, Mymensingh.

10. Hossain MI, MS Islam, FH Shikha, M Kamal and MN Islam, 2005. Physicochemical Changes in Thai Pangas (*Pangasius sutchi*) muscle during ice-storage in an insulated box. Pakistan Journal of Biological Sciences, 8: 798-804.

11. Howgate PAJ and. KJ Whittle, 1992. Multilingual Guide to EC Freshness Grades for Fishery Products. Torry Research Station, Food safety Directorate, Ministry of Agriculture, Fisheries and Food, Aberdeen, Scotland.

12. Jiang ST, DC Hwang and CS Chen, 1988. Effect of storage temperature on the formation of disulfide and denaturation of milkfish actomyosin (*Chanos chanos*). Journal of Food Science, 53: 1333-1335.

13. Konagaya S and T Konagaya, 1979. Acid denaturation of myofibrillar protein as the main cause of formation of Yake-Niku, a spontaneously done meat, in red meat fish. Nippon Suisan Gakkaishi, 45: 145.

14. Kramer DE and MD Peters, 1981. Effect of pH and refreezing treatment on the texture of yellowtail rockfish (*Sebastes flavidus*) as measured by Ottawa texture measuring system. Journal of Food Technology, 16: 493-504.

15. Mazorra-Manzano MA, R Pacgeco-Aguilear, El Diaz-Rojas and ME Lugo-Sanchez, 2000. Postmortem changes in black skipjack muscle during storage in ice. Journal of Food Science, 65: 774-779.

16. Onibi GE, Scaife JR, Fletcher TC and Houlihan DF, 1996. Influence of α-tocopherol acetate in high lipid diets on quality of refrigerated Atlantic salmon (*Salmo salar*) fillets. In: Proceedings of the Conference of IIR Commission C2, Refrigeration and Aquaculture, 20–22 March. International Institute of Refrigeration, Paris, France. pp. 145–152.

17. Pacheco-Aguilar R, ME Lugo-Sanchez and MR Robles-Burgueno, 2000. Postmortem, biochemical and functional characteristics of Monterey sardine muscle at 0° C. Journal of Food Science, 65: 2586-2590.

18. Perry SV and TC Grey, 1956. A study of the effects of substrate concentration and certain relaxing factors on the magnesium activated myofibrillar adenosinetriphosphatase. Journal of Biochemical, 64: 184-192.

19. Pirini M, PP Gatta, S Testi, G Trigari and PG Monetti, 2000. Effects of refrigerated storage on muscle lipid quality of sea bass (*Dicentrarchus labrax*) fed on diets containing different levels of vitamin E. Journal of Food Chemistry, 68: 289–293.

20. Reza MS, MAJ Bapary, CT Ahasan, MN Islam and M Kamal, 2009. Shelf life of several marine fish species of Bangladesh during ice storage. International Journal of Food Science and Technology, 44: 1485–1494.

21. Ruff N, RD Fitzgerald, TF Cross and JP Kerry, 2002. Comparative composition and shelf-life of fillets of wild and cultured turbot (*Scophthalmus maximus*) and Atlantic halibut (*Hippoglossus hippoglossus*). Aquaculture International, 10: 241–256.

22. Seki N, M Ikeda and N Narita, 1979. Changes in ATPase activities of carp myofibrillar proteins during ice storage. Nippon Suisan Gakkaishi, 45: 791-798.

23. Seki N, Y Oogane and T Watanabe, 1980. Changes in ATPase activity and other properties of sardine myofibrillar proteins during ice storage. Nippon Suisan Gakkaishi, 46: 607–615.

24. Taliadourous D, V Papadopoulos, E Domvridou, IN Sawaidis and NG Kontominas, 2003. Microbiological, chemical and sensory changes of whole and filleted Mediterranean aquacultured sea bass (*Dicentrachus labrax*) stored in ice. Journal of the Science of Food and Agriculture, 83: 1373-1379.

25. Watabe S, M Kamal and K Hashimoto, 1991. Post mortem changes in ATP, creatine phosphate and lactate in sardine muscle. Journal of Food Science, 56: 151154.

5

HAEMATHOLOGICAL AND HISTOLOGICAL EVALUATION OF AFRICAN CATFISH *Clarias gariepinus* FOLLOWING ACUTE EXPOSURE TO METHANOLIC EXTRACT OF *Khaya senegalensis*

Matouke Matouke Moise[1*] and Obadiah Audu Abui[2]

[1]Department of Biological Sciences, Federal University, Dutsinma, Katsina, Nigeria;
[2]Department of Biological Sciences, Federal Science and Technical College, Kafanchan, Kaduna State, Nigeria

*Corresponding author: Matouke Matouke Moise, E-mail: mosesmatouke@yahoo.fr

---

ARTICLE INFO   **ABSTRACT**

Haematological and histopatholoical effects of methanolic extract of *K. senegalensis* leaves was investigated on *Clarias gariepinus* over a period of 96h exposure. The median lethal concentration of the extract was 199.69mg/L. The extract caused decreased in total erythrocytes (TEC) and Packed Cell volume (PCV) respectively and increased of total leukocytes (TLC). Histopathological lesions in the liver, cytoplasmic degeneration, less intracellular space, mild necrosis, sinusoidal blood congestion and marked blood congestion in hepathocytes were recorded. However the severity but not the type of lesions was concentration-dependent. Though, the degree of tissue change (DTC) varied with the methonolic extract used. There was significant association (P<0.05) between the DTC and *K. senegalensis* methanolic concentration. The cumulative DTC indicated a moderate damage in the liver. The extract was considered toxic to the exposed fish and therefore deleterious on the organs of *C. gariepinus*.

Key words
Haematology
Histopathology
Clarias gariepinus
Methanol extract

INTRODUCTION

Khaya senegalensis is commonly used in Africa for shade, medicinal purposes and belong to the Meliacea family (Adanhounsode, 2012). Due to anthropogenic activities the plant stem and leaves might be found in the aquatic ecosystem which could be a serious threat for aquatic fauna including *Clarias gariepinus*. Exposure of aquatic organism to plant extract has been shown to have detrimental effects on fish physiology sometimes leading to mortality (Abalaka, 2015).Fish species are permanently exposed to their ecosystem and thus, exposure to a peculiar compound can act as environmental indicator. Hematology and histhopathology have been used as a biomarker for the effects of various anthropogenic pollutants on fish (Olurin *et al.,* 2006; Devi and Mishra, 2013).

Studies showed the impact of haematology and gill pathology of *Heterobranchus bidorsalis* exposed to sub lethal concentration of *Moringa oleifera* leaf extract. Erythrocytes was 2.36 ×106/mm^3, Packed Cell Volume was 20%, Haemoglobin was 6.68g/dl, mean cell volume was 85.1ft, cell haemoglobin concentration was 33.33 and mean cell concentration was 28.8pg while the mean value for white blood cell was 40.5mm^3. These results suggest a variation of values when compared to control (Olufayo and Olufunke, 2012). Many other studies showed similar variation of haematological parameters (Mammanet al., 2013; Ochang, 2007; Odeyemoet al., 2010).However, *Oreochromis niloticus* showed hispathological responses while exposed to waterborne copper, the gills were affected with edema, lifting of lamellar epithelia and an intense vasodilatation of the lamellar vascular axis. In the liver, the number of hepatocytes nucleus per mm^2 of hepatic tissue decreased with the increase of copper concentration (Figueiredo-Fernandes, 2007). In india, metal uptake by *Oreochromis mossambicus* inhabiting Indus River led to histological changes in gills and liver. The abnormalities in gills were desquamation of lamellar epithelium, hypertrophy of epithelial cells, lifting up lamellar epithelium, intraepithelial oedema, aneurysm, hyperplasis and haemorrhage. Histology of liver revealed the presence of heterogenous parenchyma characterized by vacuolization, foci of necrosis, hypertrophy of nuclei and degenerated hepatocytes (Jabeen and Chaudhry, 2013).

Intensive research on the environmental impact of chemical compound in aquatic ecosystem and also the effects of some stem bark and leaves of some plant with prominent medicinal purposes including *K. senegalensis* have been documented. However, a dearth of information about the methanolic extract of the leaves on aquatic fauna including *C. gariepinus* still exists. The study therefore aimed to evaluate the haematological indices and histopathological changes in the liver of *Clarias gariepinus* exposed to acute concentration of *Khaya senegalensis*.

MATERIALS AND METHODS

Plant extraction

K. senegalensis leaves were dried in an open airy laboratory for 2weeks and later and pounded into powder using ceramic mortar and pestle and later sieved through 100 mm sieve to obtain 500g of fine powder. 200g of the powder was packed into a soxhlet extractor with 5 L of 25% v/v methanol (98% vol. Sigma-Aldrich^R Inc., St. Louis, MO 63178, USA) as the extracting solvent. The set up was placed over water bath-bath (40- 45 °C) for 3 to 4 h for drying the substrate.

Acute fish toxicity bioassay

Three (300) Juveniles *C. gariepinus* of mean weight 18.47 ± 3.06g and standard length 10.9 ± 2.7cm. Fish were acclimatized in 500L plastic tank pond for 14 days under natural day and night photoperiods (12/12h) prior to commencement of the toxicity bioassay. Pond water was changed once every three (3) days. Fish were fed twice daily with standard feed for aquaculture.

Three hundred and sixty (360) healthy acclimatized fishes were randomly selected and distributed into 12 glass aquaria each containing 20 litres of dechlorinated water of 10 fish per aquaria, two of the aquaria served as control for acute toxicity bioassay. Twenty gram of the methanolic extract of *K. senegalensis* was obtained and dissolved in distilled water (1L) to form a stock solution of 250mg/L. 0 (control), 150, 170, 190, 210 and 230mg/L were dispersed into the experimental aquaria. The mixture was allowed to stand for about five minutes to evenly distribute via diffusion before introducing the fish. The exposure was carried out in triplicate.

At the end of the 96h exposure period, two fish were randomly sampled from the control and each treatment tanks, dissected to extract the liver.

The concentrations were selected after conducting a preliminary study, was used as an endpoint of toxicity. Probit analysis method was used to determine the median lethal concentration (LC$_{50}$ of the extract of the exposed fish. The temperature, PH, TDS, Conductivity of fish culture water was ascertained using a Hana portable hand instrument HI 98129 and the dissolved oxygen contents were measured using Winkler-Azide method and reported elsewhere (Abui and Matouke, 2015).

Haematological analyses

Total Erythrocytes (TEC) and total leucocytes (TLC) count were evaluated using an improve Neubaeur counting chamber under microscope (Dacei and Lewis, 2001). PCV was evaluated by Wintrobe method (Blaxhall and Diasley, 1971).

Histopathological analyses

At the end of the 96h exposure period, two fish were randomly sampled from the control and each treatment tanks, dissected to extract the tissues (liver). The liver was preserved in 10% formalin, washed in running tap water to remove traces of formalin, followed by dehydration using progressive percentage of alcohol and chloroform. Samples were processed, sectioned (5μm) and stained with haematoxylin and eosin using histological techniques (Bancroft and Cook 1994). Permanent slides were prepared and photomicrographs taken, using a Carl Zeiss (Axioskope 40) Trinocular Photo micrographic microscope with digital camera for comparison with tissues obtained from those of control.

The degree of tissue change (DTC), which is based on the severity of the lesions according to the methodology described by Poleksic and Mitrovic-Tutundzik (1994). For the calculation of DTC, the alteration in the liver was classified in progressive stages of tissue damage. First stage lesions (I) are slight and would be reversible with an improvement in the environmental conditions; second-stage lesions (II) are more severe, leading to effects on tissue function; and third-stage lesions (III) are very severe, with irreparable damage. The sum of the number of lesion types within each of the three stages multiplied by the stage coefficient represents the numerical value of the DTC, based on the formula DTC = $(10^0 \Sigma I) + (10^1 \Sigma II) + (10^2 \Sigma III)$, in which I, II and III correspond to the sum of the number of alterations found in stages I, II and III, respectively. The DTC was obtained for the fish of all the experimental groups and used in the statistical analysis to compare the mean degree of tissue damage between groups.

Statistical analyses

Data were analyzed using XLSTAT software version 15.5. Means (SD) were subjected to ANOVA and Chi-square for statistical significance (P< 0.05).

RESULTS

Quality of the water

The physicochemical parameters of water showed that EC was 79.5μs/cm, TDS 56.79 mg/L, PH 6.92 and Temperature 24.14 °C (Table 1).

Table 1. Physicochemical parameters of fish culture water of *C. gariepinus* exposed to methanol extract of leaves *K. senegalensis*.

Physicochemical parameters	Value
Conductivity (5μs/cm)	79.5±3.48
Total dissolved solid (TDS) mg/L	56.79±1.81
PH	6.92±0.12
Dissolved oxygen (DO) (mgO$_2$/L)	6.66±0.14
Temperature (^{0}C)	24.14±0.18

Acute fish toxicity bioassay

The effect of *K. senegalensis* leaves on *C. gariepinus* showed a high mortality of 60, 83 and 70 at concentration 210 and 230 mg/l respectively. The mortality increased significantly with the increased in concentration (P< 0.05). The median lethal concentration (LC50) of leaves extract was therefore 199.69 mg/L (Table 2).

Table 2. Mortality in *C. garipinus* exposed to methanol extract of *K. senegalensis* leaves over 96h period.

Extract concentration (mg.L)	Log concentration	Total Mortality	Percentage total mortality (%)	Probit value
0.0 (control)	0.000	0.0	0.0*	2.5*
150mg.L	2.176	1	10	3.72
170mg.L	2.230	3	30	4.16
190mg.L	2.279	4	40	4.75
210 mg.L	2.322	6	60	5.25
230 mg.L	2.362	7	70	5.52

Y= 9.33x-21.47, LC_{50}= 199.69, * =corrected value

The surface of *C. gariepinus* liver in control group had a preserved architecture, normal hepatocytes and absence of blood congestion while compared to other concentration (Figure 1, 2, 3, 4, 5, 6 and 7).

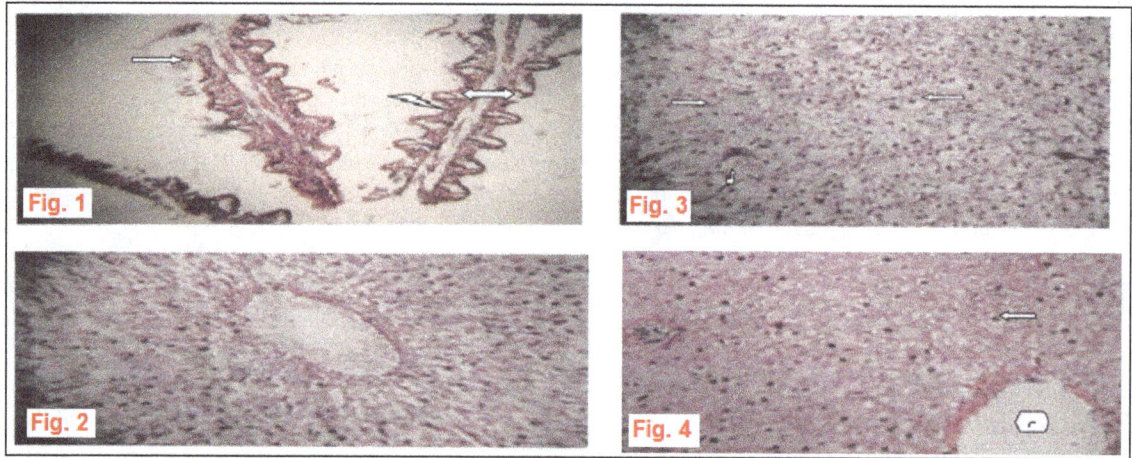

Figure 1. Photomicrograph of the liver of *C. gariepinus* showing a preserved architecture and normal hepatocytes with sinusoids intact; **Figure 2.** Photomicrograph of the liver of *C. gariepinus* showing a preserved architecture and necrosis of the hepatocytes at 170mg/L of toxicant; **Figure 3.** Photomicrograph of the liver of *C. gariepinus* showing mild necrosis of the hepatocytes at 190mg/L of toxicant; and Figure **4.** Photomicrograph of liver of *C. gariepinus* showing marked necrosis of the hepatocytes at 210mg/L of toxicant.

Haematological evaluation

The haematological evaluation showed the highest means total erythrocyte concentration (TEC) (238±3.0) at concentration 150mg/L and the lowest (122.33±4.16) at concentration 230mg/L. The total leucocytes concentration in the blood sample was highest (2888±16) and the lowest 1147.23 at control level. The pack cell volume (PCV) was the highest (39.13) during control and the lowest 13.67±1.53 at concentration 230mg/L (Table 3). There was no significant difference among means of haematological parameters.

Table 3. Means of haematological parameters on *Clarias gariepinus* to acute concentration of methanol leaf extract of *Khaya senegalensis*.

Concentration mg/L	TECx10^6	TLCx500 m^3	PCV
0	252^a	1147.23^c	39.13^a
150	238±3.0^b	1626.67±16.65^d	31±2.0^a
170	212.0±2.65^c	2072 ±8.0^c	23.67±2.51^b
190	201±3.0^d	2422.67±28.4^b	21.33±3.06^b
210	159.33±1.53^e	2870.67±22.03^a	13.33±3.06^c
230	122.33±4.16^f	2888. ±16^a	13.67±1.53^c

Means with the same letters along the columns are not significantly different (P>0.05).

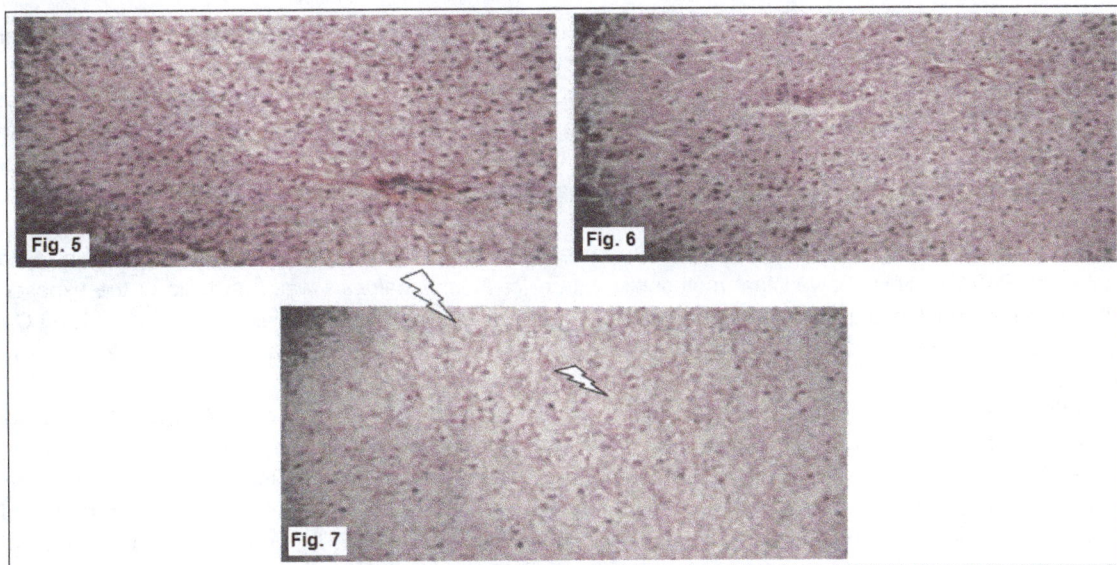

Figure 5. Photomicrograph of the liver of *C. gariepinus* showing marked necrosis of hepatocytes at 230mg/L of toxicant; **Figure 6.** Photomicrograph of the liver of *C. gariepinus* showing marked necrosis of hepathocytes at 230 mg/L of toxicant; and **Figure 7.** Photomiccrograph of the liver of *C. ganepius* showing marked necrosis of hepatocytes at 230 mg/L of toxicant.

Table 4. Type and incidence of histological lesion in the liver of *Clarias gariepinus* exposed to methanol extract of *Khayasenegalensis* leaves over 96h periods.

Stage	Degree of tissue damage in the Liver
I	Degeneration of the cytoplasm(++)
	Less intracellular space (++)
II	Mild necrosis (++)
	Sinusoidal blood congestion (+++)
III	Marked blood congestion in hepathocyte (++)

(+): low incidence (++): moderate incidence   (+++): high incidence

The degree of tissue change (DTC) was highest (111.67) when the fish was exposed to 230mg/L concentration, this translate irreversible damage observed. The DTC of 72 which is moderate was observed when the fish was exposed to 210mg/L while the concentration less than 210mg/L was without danger for the liver of the fish (Table 5). The result showed significant difference when compared to the control ($P<0.05$). The cumulative degree of tissue change showed a moderate change (55.06±12.7) in the liver of fish (Table 6).

Table 5. Degree of tissue change per extract concentration in the liver of *Clarias gariepinus* exposed to methanol extract of *Khaya senegalensis* leaves over 96h period.

Concentration	00	150	170	190	210	230
Liver	12±8.5	53± 9	42.67±30.2	39±11	72± 11	111.67± 6.6

DISCUSSION

The physical and chemical parameters analyzed in this study are very relevant for the status of water quality tolerable by *C. gariepinus*. The result of the physical chemical in this study, conductivity 79.5 µs.cm, TDS 56.79mg/L, Temperature 24.14°C PH 6.92 and DO 6.66mgO$_2$ 132mg/L corroborate with the standard of freshwater fish water quality. The physicochemical parameters of fish culture water in this study were all within acceptable limit for the survival of *C. gariepinus*. The standard as stated by Kalawole *et al.* (2011) reported a conductivity less than 135-100µs/cm, DO greater than 4 mg/L; PH between 6.5 to 9 and Temperature between 20-30 °C.

In this study the mortality increased with the increase of concentration, this increase was shown to be significant ($P<0.05$). This showed that methanolic extract of *K. senegalensis* was 138 toxic to the exposed fish. The dependence of plants extract to mortality of *C. gariepinus* was similar Abalaka *et al.* (2015). The LC$_{50}$ value of 199.69 in this study was higher compared to Ayuba *et al.* (2012) who had a LC50 of 120.23 mg/L while studying the acute toxicity *C. gariepinus* exposed to *Daturainnoxia* leaf extract.

A decrease of total erythrocytes observed in this study when compared to the control suggests a gradual damage or inhibition of oxygen production of erythrocytes in the blood of fish. The increased and decreased values of TLC suggest some abnormalities. Leukocytes are white colored blood cells which defend the animal body against infections and diseases. They might increase in number in this study to boost the immunity of fish against certain infections due to variation of concentration extract. The PCV measures the percentage of red blood cells (RBC) to the total blood volume. These values obtained in this study decreased respectively with the increase of concentration which suggests an increase of anemia in fish. This study agreed with finding on the haematological responses of *Clarias gariepinus* exposed to sublethal concentration of *Daturainnoxia* root extract (Okomodaetal, 2013). *K. senegalesis* could be suspected to cause gradual anemia in fish through hemolysis though no significant difference was observed in the measured haematological parameters.

Using the methods described by Polekesic and Mitrovic-Tutundsie (1994) as reference, five (5) types of lesions were identified in the liver of *C. gariepinus* (Table 4). Two (2) of which were first stage (I), two (2) were second stage (II) and one (1) was stage three (III).

The liver is the organ most associated with detoxification and accumulation due to the function and blood supply, it is also affected by contaminant in the ecosystem and also play a critical role in fish physiology because acts as the main storage for many substance (Moneim-Abdel *et al.*, 2008). Histopathological changes of liver in this study gave an insight on the nature of the stressor *K. senegalensis* leading to pathological changes in tissues. In this study the damage in the liver of *C. gariepinus* exposed to various concentration accounts for the response to the methanol extract *K. senegalensis* which act as a toxicant. The response in the liver tissues showed a degeneration of the cytoplasm, less intracellular, mild necrosis, sinusoidal blood congestion and marked blood congestion in hepathocytes. These changes may be attributed to direct toxic effects of pollutants on hepatocytes as found in *K. senegalensis*. The observation in this study suggests a structural damage of the hepatocytic cytoplasm with gradual increase of *K. senegalensis* concentration. Elsewhere, authors described hepathocytic alterations, hypertrophy of hepatocytes, and infiltration of leukocytes, necrosis and fibrosis in *Channapunctatus* a freshwater fish after exposure to pesticide chlorpyrifos (Devi and Mishra, 2013).

Less intracellular spaces, mild necrosis, sinusoidal congestion and marked blood congestion in this study might be the cause of the degeneration of hepathocytes with cytoplasmic vacualation and hypertrophy. These observations were reported in *Poeciliareticulate* exposed to acute concentration of textile effluent (Selvaraj *et al.*, 2015). Increased vacualisation of the hepathocytes could be described as a signal of degenerative process. Cytoplasmic degeneration in this study suggests metabolic damage, probably related to exposure to acute *K. senegalensis*.

Marked blood congestion and marked necrosis lesions belonging to stage I and stage II respectively observed in this study were in agreement with many studies that examined the effect of different pollutant on fish liver, though the pollutant were metals (Olurin *et al.*, 2006, Figueiredo-Fernandes *et al.*, 2007; Devi and Mishra, 2013; Jabeen and Chaudhry, 2013).

The highest mean DTC was observed at concentration 230mg/L might suggest a severe damage in the hepatic cells if prolonged. But the cumulative mean of 55.06 was observed in the liver of fish exposed to acute concentration. This result agreed with findings of (Carmago and Martinez, 2007) which report showed a mean DTC of 52.34 indicating that in most cases the hepatic lesions caused moderate damage to tissues though, an increase of DTC 174 in summer was reported while studying the histopathology of liver of a Neotropical fish caged in an urban stream. Studies conducted on the gills and skin of *C. gariepinus* exposed to acute concentration of ethanol extract of *Adenium obesum*stem bark revealed a cumulative DTC of 23.12 and 1.78 in the gills and skin respectively which suggest a low damage while compared to the present study.

CONCLUSION

In the present study, haematological parameters (TEC TLC and PCV) and histopatological changes have been related to acute concentration of *K. senegalensis*. It can be concluded that haematological indices and liver alteration of fish may serve as biomarker of toxicity due to acute concentration of *K. senegalensis*. However, other studies are necessary to evaluate de degree of tissue change in other species of fish.

CONFLICT OF INTEREST

The authors have no conflicts of interest concerning the work reported in this paper.

REFERENCES

1. Abalaka SE, Muhammad, YF, Najume, DG and SF Ambali, 2015. Gills and skin histopatholigicalevaluation in African Shaptooth catfish, *Clarias gariepinus* exposed to ethanol extract of *Adenium obesum* stem bark. Egyptian Journal of Aquatic Research, 41: 119-127.

2. Abui OA and MM Matouke, 2015. Toxicological evaluation of methanol extract of Khaya *senegalensis* leaves in African catfish *Clarias gariepinus*. Journal of Applied Science and Environmental Management, 19: 557-560.

3. Adanhounsode SN, 2012. Problematique de l'utilisation de *khayas enegalensis* comme arbre de reboisement urbain cas de la ville de cotonou. These publiee pour l'obtention du diplome de licence professionnelle. Ecole Polytechnique d'Abomey Calavi. www.memoireen ligne.com.

4. Ayuba VO, Ofojekwu PC and SO Musa, 2012. Haematological response and weight changes of the African Catfish *Clarias gariepinus* exposed to Sub-lethal concentration of *Daturainnoxia* root extract. Production and Agriculture Technology, 8: 134-143.

5. Bancroft JD and HC Cook, 1994. Manual of Histological Techniques and their Diagnostic Application. Churchill Livingstone, London, pp. 289-305.

6. Blaxhall PC and Diasley KW, 1973. Routine haematological methods for use with fish blood. Journal of Biology, 5: 771-781.

7. Carmago MM and C Martinez, 2007. Histopathology of gills, kidney and liver of a Neotropical fish caged in an urban stram. Neotropical Ichthyology, 5: 327-336.

8. Dacei JW and SM Lewis, 2001. Practical haematology, 9thed Churchill, Livingston, London.

9. Figueiredo-Fernandes A, Ferreira-Cardoso JV and S Garcia-Santos, 2007. Histopathological changes in liver and gill epithelium of Nile tilapia, *Oreochromis niloticus*, exposed to water borne copper. Prequisa. Veterina. Brasileira, 27: 103-109.

10. Jabeen Farbeen A and AS Chauhry, 2013. Metal uptake and histological changes in gills and liver of Oreochromis mossambicua inhabiting Indus River. Paskitan Journal of Zoology, 45: 9-18.

11. Kalawole AE, Olusegun AO and IA Ayodele, 2011. Fundamentals of fish farming in Nigeria. Published and printed by Walecrown ventures Ibadan, Nigeria. Pp. 40-42.

12. Mamman T, Ipinjolu JK and I Magawata, 2013. Haematological indices of *Clarias gariepinus* (Burchell, 1882) fingerling fed diet containing graded level of Calabash (*Lagenaria vulgaris*) seed meal. Journal of Biology, Agriculture and Healthcare, 3: 100-104.

13. Mishra P and S Gupta, 2014. Haematological evaluation of *Eclipta alba* root extract in Catfish, *Clarias bratrachus* (Linnaeus, 1758). Journal of Pharmaceutical and Scientific Innovation, 3: 239-244.

14. Monein-Abdel A, Shabana, Khadre EM and HH Abdel-Kader, 2008. Physiological and histopathological effects in catfish exposed to dyestuff and chemical wastewater. International Journal of Zoology Research, 4: 189-202.

15. Ochang SN, Oyedapo F and TA Olabode, 2007. Growth performance, body composition, haematology and product quality of the African Catfish (*Clarias gariepinus*) fed diets with palm oil. Pakistan Journal of Nutrition, 6: 452-459.

16. Odeyemo OK, Adedeji OB and CC Offor, 2010. Blood lead level as biomarker of environmental lead pollution in feral and cultured African catfish (*Clariasgariepinus*), Nigerian Veterinary Journal, 31: 139-147.

17. Okomoda VT, Ataguba GA and VO Ayuba, 2013, Haematological response of *Clarias gariepinus* fingerlings exposed to acute concentration of Sunate. Journal of Stress Physiologyand Biochemistry, 9: 271-278.

18. Olufayo M and AOlufunke, 2012. Haematology and gill pathology of *Heterobranchus bidorsalis* exposed to sublethal concentration of Moringaoleifera leaf extract, Journal of Agriculture and Biodiversity Research, 1: 18-24.

19. Olurin KB, Olojo EA, Mbaka GO and AT Akindele, 2006. Histopathological responses of the gill and liver tissues of *Clarias gariepinus* fingerlings to the herbicide glyphosate. African Journal of Biotechnology, 5: 2480-2487.

20. Poleksic V and V Mitrovic-Tutundzie, 1994. Fish gills as a monitor of sublethal and chronic effects of pollution In: Mulls, R, Llyod (Eds) on freshwater fish. Fishing News Books, Oxford. London. Pp339-352.

21. Selvaraj D, Leena R and C Kamal, 2015. Toxicological and histopathological impacts of textile dyeing industry effluent on a selected teleost fish *Poecilia reticulate*. Asian Journal of Pharmacology and Toxicology, 3: 26-30.

6

STUDY ON THE SOCIO-ECONOMIC CONDITIONS OF THE FISHERMEN IN TEKNAF

Subrata Kumar Ghosh[1]*, Mirja Kaizer Ahmmed[1], Sk. Istiaque Ahmed[1], Md. Khabirul Ahsan[2] and Md. Kamal[2]

[1]Faculty of Fisheries, Chittagong Veterinary and Animal Sciences University, Chittagong-4225, Bangladesh; [2]Faculty of Fisheries, Bangladesh Agricultural University, Mymensingh-2202, Bangladesh

*Corresponding author: Subrata Kumar Ghosh, E-mail: subratacvasu@gmail.com

---

ARTICLE INFO

Key words

Socio-economic
Fishermen
Fishing
Teknaf

ABSTRACT

Study was conducted on the socio-economic conditions of the fishermen in Teknaf to evaluate fishermen livelihood and social status for a period of one year from March, 2014 to April, 2015. Data were collected in terms of income generation, age distribution, housing, literacy rate, health and sanitation facilities of the fishermen. Fishing was regarded as the major source of income of the traditional fishermen but occasionally they undertook a variety of non-fishery related activities, which constituted a substantial part of their annual income. Among 105 fishermen interviewed, 59.25% were below 30 years, 29.62% were between 30 and 39 years, and the remaining 11.11% were more than 40 years old, and their literacy level was 62.96% illiterate, 18.51% can write their names, 14.81% had received education up to primary level and 3.70% had received secondary education. Income distribution showed significant inequality between marginal and non-marginal fishermen from group fishing. The national and local NGO like BRAC, ASA provided credit only to the organized poor members to purchase fishing gears and boats but the amount of credit provided by the NGOs was insufficient and could not commensurate to the poor people's actual need. The present study revealed that sanitary conditions of the fishermen were very poor that 25% fishermen have semi-constructed and 10% of the fishermen had no sanitary facilities. Most of the fishermen (65%) have un-constructed sanitary facilities. The present study suggested that there is a clear need to legalize fishing profession and their settlement as well as traditional fishing communities should be given priority in getting necessary supports from all concerns.

INTRODUCTION

Fish and fishing business is an important sector of many nations of the world from the standpoint of income generation and employment generation. Fishing plays an important role in supporting livelihood worldwide and also forms an important source of diet for over one billion people. Fisheries sector plays an important role in the economy of Bangladesh by contributing to the national income, employment and foreign exchange. According to Rao *et al.* (1988), fish is a valuable source of protein and occupies a significant position in the socio-economical fabric of South-Asian countries.

In Bangladesh, a lot of people are engaged in fishing activities in the Teknaf region as a source of their livelihood. Teknaf coast has grown considerably over the period and playing a very important role in the local economy, employment generation, foreign exchange earnings, food security and livelihood of the local community. So, it is essential to identify and evaluate marine fishery resources with mode of utilization, which will bring sustainable developments and improve the livelihood condition of concern stakeholders in Teknaf. Taking all the aspects into account the current research activity was undertaken to study the socio-economic conditions of the fishermen in Teknaf.

MATERIALS AND METHODS

Study Area

The study was conducted in Teknaf region because an appreciable number of people are engaged in fishing activities in this area.

Data Collection

Primary data were collected by field survey that involved the investigation of the people in terms of income generation, age distribution, housing, literacy rate, health and sanitation facilities of the fishermen. Data were collected for a period of one year from March, 2014 to April, 2015 where a total of 105 fishermen were interviewed.

Data Collection Methods

Questionnaire Survey

For questionnaire and interviews, fishermen were selected through simple random sampling. A total of 105 fishermen were interviewed during the study period. Interviews were conducted at different time of the day. The interviews focused on socio-economic conditions of the fishermen in terms of income generation, age distribution, housing, literacy rate, health and sanitation.

Participatory Rural Appraisal (PRA)

PRA is a group of methods to collect information in a participatory basis from rural communities. The advantage of PRA over other methods is that it allows a wider participation of the community, the information collected is likely to be more accurate (Chambers, 1992; Nabasa *et al.*, 1995). This study used PRA tool: focus group discussion (FGD) with fishermen and associated groups. Among the 105 fishermen interviewed 70 were male and 35 were female (Figure 1). A total of 7 FGD sessions among different groups containing 5 females and 10 males were conducted in the study and duration of each FGD was approximately two hours.

Data Processing and Analysis

After the collection of data they were scrutinized and carefully edited to eliminate possible errors and inconsistencies contained in the schedules. After completing the pre-tabulation task, the primary data were entered, processed and analyzed through simple statistical methods using Microsoft excel.

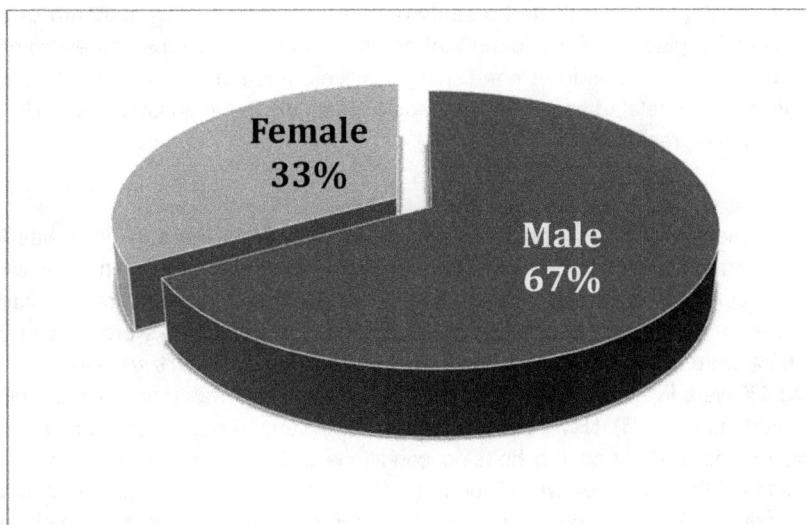

Figure 1. Percentage of sex ratio in the study area

RESULTS AND DISCUSSION

The study was conducted on the socio-economic conditions of the fishermen in Teknaf region. Emphasis was given on income and living standard, housing condition, age and family members, family size, literacy and education, occupational status, credit access issues, sanitary facilities of the fishermen.

Income and living standard:

Although fishing is the major and, in some cases, the only source of income of the traditional fishermen, the fisher folk occasionally undertake a variety of fishery related and non-fishery related activities, which constitute a substantial part of their annual income (Table 1). Their income augmenting opportunities, however, are very limited. Fishery related activities carried out in the study area include fish marketing and trading, gear and craft maintenance and repair etc. There are very limited options for non-fishery related activities such as wage labor in the other sectors like agriculture, construction, livestock and poultry raising, etc. It was found that, about 76.4% of total income comes from fish and fishery related activities and a fisherman earns about 40,000 taka per year. Rest 23.6% of total income comes from non-fishing activities and a farmer earns about 10,000 taka per year.

Table 1: Average annual income of fishermen from fishing, fishery and non- fishery related activities

Sources of income	Income of(TK) per year	% of total income
Fish and fishery related activities	30,000	76.4%
Group fishing		
Fish marketing		
Net lending		
Non –fishing activities (Agriculture)	10,000	23.6%

Mazumder et al. (2014) conducted a study on socio-economic condition of fishermen in Jelepara under Pahartoli of Chittagong district. They interviewed 50 fishermen and found that 20% fishermen's yearly income was found BDT 50000-70000, 48% fishermen's yearly income was between BDT. 71,000-1,00,000 and 32% fishermen's yearly income was found above BDT 1,00,000. They included the income coming from fishing activities only whereas the present study investigated the income coming from both fishing and non-fishing activities. Chowdhury et al. (2011) conducted a study on the income and living standard of the fishermen in the Teknaf peninsula of Bangladesh. They found that about 82% of total income comes from fish and fishery related activities and rest 18% comes from non-fishing activities. It is not very unusual to obtain most of the income sources from fisheries related sector for the people inhabiting in the coastal basin. The result is almost similar to the findings of the present study.

Housing condition

Most of the fishermen lived in very poor housing conditions. From the survey, it was found that 25% households of the fishermen were tin shed with bamboo. 60% households were tin shed with tin wall, 14% households were containing of straw components and 1% household were containing building, respectively (Figure 2). Mahmud et al. (2015) interviewed 50 fishermen and found that 14% were lived in the house made by straw and soil, 62% were lived in the house made by tin, wood and soil, 20% were lived in the house made by brick and tin and 4% were live in Building. In our study, the small size was more than double than the study conducted by Mahmud et al. (2015). Nevertheless, their findings support our results to a considerable degree. Hossain (2002) conducted a study on the housing conditions of the fishermen in Teknaf. They found that about 30% households of the fishermen were lived in tin shed with bamboo, 55% households were in tin shed with tin wall, 13% households were containing of straw components and 2% household were containing building. The results are an index of below living standards of the people involved in fishing. The documented data are still more or less in similar conditions with the data recorded 14 years ago by Hossain (2002). This indicates the living standards of the fishers of the Bay of Bengal changes nothing but decreases a bit in comparison with the previously collected data.

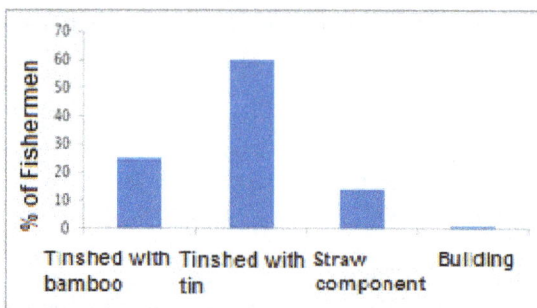

Figure 2. Housing condition of the fishermen

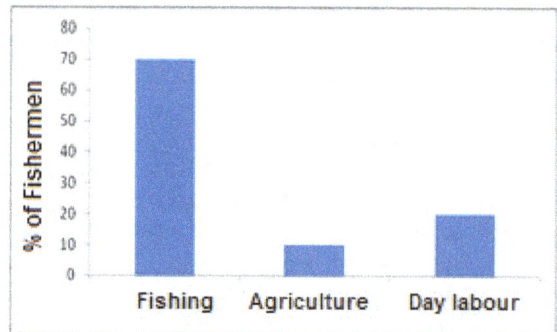

Figure 3. Occupational status of the fishermen

Age and family members

Age and family members of the fishermen were also interviewed. From the interviews it was found that 65 fishermen (59.25%) were below 30 years old, 34 fishermen (29.62%) were between 30 and 39 years, and the remaining fishermen (11.11%) were more than 40 years old (Table 2). The result indicated that the middle age groups are involved in fishing activities. Ahmed (1999) in coastal region reported 66% and 70% under 40 years age, respectively. The Bangladesh Bureau of Statistics (BBS, 1998) reported that fishing households in Bangladesh have higher family size than the national average. Mursheduzzaman (2007) reported that age group of 31-40 years was the highest (50.00%) and 41-60 years was the lowest (17.00%) and 21-30 years was 33% within considering 250 fishermen. The tendency of lower percentage in the higher age groups may be for the unavailability of advanced medical facilities as well as higher prone to the natural calamities. The recorded results here is more or less similar to our study.

Literacy and education

From the survey, 64 fishermen (62.96%) were found to be illiterate and cannot write their names, 20 fishermen (18.51%) were semi-literate who only can write their names, 17 fishermen (14.81%) had received education up to primary level and 4 fisherman (3.70%) had received secondary education, respectively (Table 3). Rahman, et al. (2012) reported that among 100 fishermen, 66.66% were illiterate while 16.66% had primary education and and another 16.66% can sign only in the Nijhum island under Hatiya upazilla of Noakhali district in Bangladesh. Inadequate consciousness and lack of educational infrastructure in the concerned area are the prime reason of the lower literacy rate. BBS, 2011 classifies literate persons as those who can write a letter in any language. Mazumder et al. (2014) found that among the 50 fishermen, there was no H.S.C and S.S.C passed people. 16% had passed class five, whereas 14% fishermen were can sign only and 70% were illiterate. The result is almost similar to the previous studies.

Table 2. Family size of the fishermen in the study area.

Family Size	% of total fishermen
Small Family(2-4)	10%
Medium Family(5-7)	60%
Large Family(8-10)	30%

Table 3. Educational status of fishermen in the study area.

Educational status	No. of fishermen (n=105)	% of total fishermen
Illiterate	64	62.96
Capable to sign only	20	18.91
Primary	17	14.81
Secondary	4	3.70

Occupational status

The present study has also revealed that 70% of fishermen are engaged in fishing as their main occupation, 10% was in agriculture and 20% in daily laborer as in other business (Figure 3). Mahmud et al. (2015) that the main occupation of the people among 50 fishermen were fishing (86%), while 14% were secondary occupation. This result supports our findings to an acceptable level. Haque (2008) described the selected characteristics of coastal fishermen and their participation in alternative livelihood activities. Alternative livelihood activities were crop cultivation, vegetables cultivation and fruits cultivation in field/homestead and reported that 80% of fishermen were engaged in fishing as their main occupation, and rest 20% of fishermen engaged in non-fishing activities. As the study area supports a large scale of marine fisheries of the country, fishing, unsurprisingly the primary occupation for the adjacent people.

Sanitary Facilities

The present study revealed that sanitary conditions of the fishermen were very poor that 25% fishermen have semi-constructed and 10% of the fishermen had no sanitary facilities. Most of the fishermen (65%) have un-constructed sanitary facilities (Figure 4). Mahmud et al. (2015) found that among the 50 fishermen, on an average 13 (26%) household used unconstructed sanitary facilities, 26 (52%) household used semi-constructed sanitary facilities and only 3 (6%) household used constructed sanitary facilities. But 8 (16%) had no sanitary facilities that they use agricultural land, crop field, canals, bushy area and hidden place. The result indicated that there is a greater degree of similarities with our findings. Rahman et al., (2012) reported that 95% households used un-constructed sanitary facilities in the Nijhum Island under Hatiya upazilla of Noakhali district in Bangladesh. The poor income of the concerned fishers in the present study area mainly led them to build such unhygienic sanitary facilities. Besides lack in the common people motivation is another vital reason for such conditions in the study period.

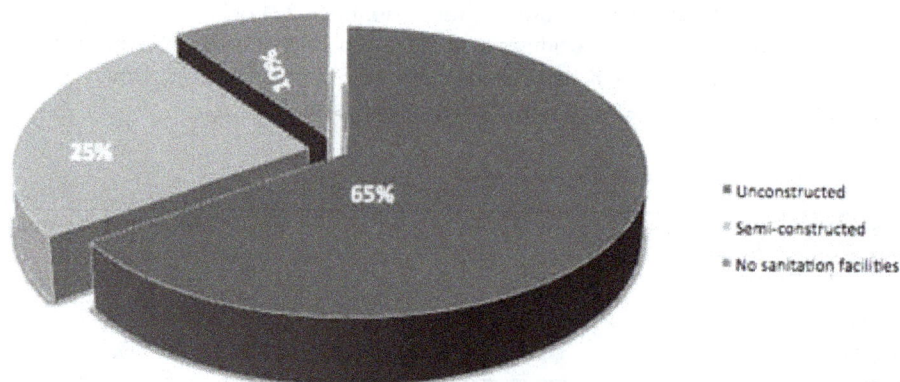

Figure 4. Sanitary facilities of the fishermen in the study area

Credit access issues

The national and local NGO like BRAC, ASA provide credit only to the organized poor members for purchase fishing gears and boats. They have taken loan at 15% interest from ASA and BRAC, 12% interest from landlord. This interest rate varies from season to season. It is often argued that the amount of credit being provided by the NGOs is insufficient and is not commensurate to the poor people's actual need. After repayment only 36% became self-sufficient who did not need financial help but 12% borrow money from their neighbors, 15% from relatives, 30% from NGOs and 7% from co-operatives for their fishing business (Figure 5).

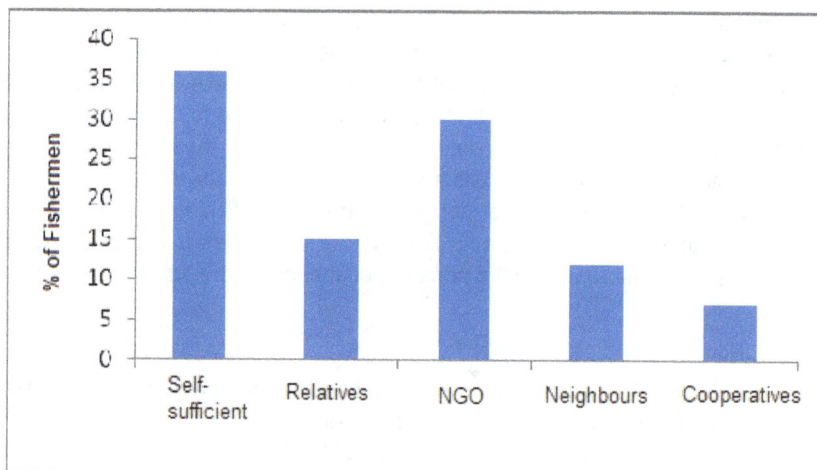

Figure 5. Sources of credit facilities of the fishermen

Constraints in the fishing community

Socio economic constraints such as household family pressure, low income, illiteracy, low economic status and lack of alternative employment opportunities are the main problems for marine fisheries development. The offered credit facilities from different GOs, NGOs are insufficient to meet their needs. Besides, in receiving such credit facilities they need to pay high interest. These socio economic factors are affecting marine resources. Fishermen are also facing problems on child education, nutrition, cooking fuel, animal feed and house building materials. Almost all fishermen mentioned lack of capital and the increasing fishing pressure as their main problems. The fishermen of Bangladesh are socially disadvantaged and lacking in fulfilling their basic needs (DFID, 1998). According to Rahman (1994), fishermen were below the poverty line and were struggling to survive, with health, nutrition, sanitation, water supply, soil fertility, cooking fuel, animal feed and house building materials as their day-to-day problems.

CONCLUSION

The implication of this study is that the socio-economic conditions of fishermen in Teknaf were not satisfactory. The fishermen were deprived of many amenities. Fish production and fish fauna of the area was being drastically reduced due to environmental and manmade activities such as over-fishing, using restricted gear indiscriminate use of fishing gears and as a whole due to absence of management policy. In the event of decreasing fish resources in Teknaf peninsula, supplementary income from other than fishery is of great importance. Goat and sheep rearing, cow rearing, home based vegetables, agriculture, small business etc. are identified as important alternative income generating sources for male whereas, handicraft, net making and duck and poultry rearing are especially identified for women. So, necessary step should be taken by GOs and NGOs to assist the fishermen to adopt these alternative income-generating sources.

REFERENCES

1. Ahmed NU, 1999. A study on socio-economic aspect of coastal fishermen in Bangladesh. MS Thesis, Department of Aquaculture, BAU, Mymensingh, Bangladesh.
2. BBS, 1998. Statistical Year book of Bangladesh, 1997, Bangladesh Bureau of Statistics, Statistics Division, Ministry of Planning, Government of the People's Republic of Bangladesh, Dhaka, Bangladesh.
3. BBS, 2011. Statically yearbook of Bangladesh, Bangladesh Bureau of Statistics, Statistical division, Government of the peoples of Republic of Bangladesh, Dhaka. 580p.
4. Chambers R, 1992. Rural Appraisal: Rapid, Relaxed and Participatory, IDS Discussion Paper No. 311, Institute of Development Studies (IDS), Brighton, UK.
5. Chowdhury MSN, Hossain MS, Mitra A, and Barua P, 2011. Environmental functions of the Teknaf Peninsula mangroves of Bangladesh to communicate the values of goods and services. Mesopotamian Journal of Marine Science, 26: 79-97.
6. DFID, 1998. Sustainable livelihoods guidance sheets, Department for International Development (DFID), London, UK.
7. Haque HA, 2008. Knowledge of coastal communities on biodiversity conservation issues, MS Thesis, Department of Aquaculture, BAU, Mymensingh, 36-38.
8. Hossain MS, 2002. Artisanal Fisheries Status and Sustainable Management Options in Teknaf, 8-18pp.
9. Mahmud S, Ali ML and Ali MM, 2015. Present scenario on livelihood status of the fishermen in the paira river, southern Bangladesh: constraints and recommendation. International Journal of Fisheries and Aquatic Studies, 2: 23-30.
10. Mazumder SK, Hossain S, Hasan MT and Alam MT, 2014. Socio-Economic Condition of Fishermen in Jelepara under Pahartoli of Chittagong District, 1: 65-72
11. Mursheduzzaman M, 2007. Socio-economic study of the fishermen community of Ichamatiriver in Santhiaupazilla under Pabna district, MS Thesis, Department of Aquaculture, BAU, Mymensingh, 76.
12. Nabasa J, Rutwara G, Walker F and Were C, 1995. Participatory Rural Appraisal: Practical Experience. Natural Resources Institute (NRI), Greenwich University, London.
13. Rahman M, Rahman MM, Hasan MM and Islam MR, 2012. Livelihood status and the potential of alternating income generating activities of fishers community of nijhum dwip under hatiyaupazila of Noakhalidistric in Bangladesh. Bangladesh Research Publiction Journal, 6: 370-379.
14. Rahman AKA, 1994. Country report on socio-economic issues in coastal fisheries management in Bangladesh. In: Socio-economic Issues in Coastal Fisheries Management, Proceedings of IPFC Symposium, Bangkok, Thailand, 23-26 November, 1993. FAO Indo-Pacific Fishery Commission, No. 8. 170-175.
15. Rao LM, Rao GV and Sivani G, 1988. Hydrobiology and Ichthyofauna of Mebadrigedda Stream of Vishakhapatnam, Andhra Pradesh. Journal of aquatic Biology, 13: 1-2.

7

IMPACT OF FIFTEEN DAYS FISHING BAN IN THE MAJOR SPAWNING GROUNDS OF HILSA (*Tenualosa ilisha* HAMILTON 1822) ON ITS SPAWNING SUCCESS

Md. Anisur Rahman, Flura*, Tayfa Ahmed, Md. Mehedi Hasan Pramanik and Mohammad Ashraful Alam

Bangladesh Fisheries Research Institute, Riverine Station, Chandpur-3602, Bangladesh

*Corresponding author: Flura; E-mail: flura_bfri@yahoo.com

ARTICLE INFO

ABSTRACT

Key words

Impact
Hilsa
Spawning
Ban

The present study was conducted to assess the impact of fifteen days fishing ban on breeding success of hilsa shad in the major spawning grounds of hilsa in the month of September and October, 2015. The study showed that fishing ban during spawning seasons have significant role in the successful reproduction of hilsa. In and around the spawning grounds among all the captured hilsa, male: female ratio was found 1: 1.86 and percent composition was 35% and 65% respectively. In 2015, percent composition of spent hilsa during major breeding period in the spawning grounds was found 36.60%. The eggs production of hilsa was calculated, in the year 2010, 2011, 2012, 2013, 2014 and 2015 about 336199Kg, 385500Kg, 380400Kg, 447100Kg, 417765Kg and 494365Kg respectively, hilsa eggs could have been produced indicating a positive impact of 15 days fishing ban in the spawning season. Comparatively higher percentages of gravid hilsa were found which were not available in the similar quantity and condition in the fishing ban period in other than spawning areas of hilsa. During the present investigation, fairly higher amount of spent hilsa and juveniles were observed in the spawning grounds. On the other hand, fewer juveniles and spent hilsa were observed in the adjacent areas of the spawning grounds indicating that in comparison to the recent reports there might have little or no changes of the spawning grounds of hilsa occurred. Along with the jatka fry, spawn and fries of other fishes were also found in higher quantity than the previous years and thus it is assumed that 15 days fishing ban also might have positive impact on the successful breeding of other fishes. Overall, the fishing ban was found effective for successful breeding of hilsa.

INTRODUCTION

The hilsa shad, commonly known as Hilsa (*Tenualosa ilisha*, Hamilton 1822) referred in the literature as an anadromous (earlier) Clupeid of the Bay of Bengal and Indian Ocean, now established as a diadromous ascends in the rivers flowing into the Bay of Bengal, Arabian Sea and Persian Gulf. Hilsa belongs to the subfamily Alosinae, family Clupeidae, order Clupeiformes, and is one of the most important tropical fishes of the Indo-Pacific region. It is a fast swimming euryhaline known for its cosmopolitan distribution in brackish water estuaries and marine environment. Naturally hilsa is in great demand globally, specifically in the oriental world and enjoys high consumer preference. Its high commercial demand makes it a good forex earner. This is an important migratory species in the Indo-Pak sub-continent, especially in Bangladesh, India and Myanmar. *Tenualosa ilisha* (Fisher and Bianchi, 1984) is the most widespread tropical shads found from north Sumatra in the east to Kuwait in the west and is the basis of important fisheries in Bangladesh, India, Burma, Pakistan and Kuwait (Al-baz and Grove, 1995; Whitehead, 1985; Blaber, 2000). It is the national fish of Bangladesh and the largest single species fishery contributing 75% of total catch in this region (Raja, 1985) that accounts nearly half of the total marine catch and about 12-13% of total fish production of the country (Haldar, 2008).

Impact of fifteen days fishing ban

The hilsa fishery was declining tremendously over the last decades for increasing fishing pressure and environmental degradation from the inland open water although the total marine production remains more or less static. In an investigation, Haldar and Rahman (1998) found that hilsa landing at Chandpur (a major landing center) has lost about 25.8% from 1978-88 to 1989-94 due to loss of freshwater discharge from the upstream international river. Construction of cross dam and flood control dam has destroyed a commercial hilsa fishery of about 500 MT/yr (Haldar et al., 1992). In addition, during the spawning migration (marine water to freshwater and vice versa) large numbers of gravid and immature fishes are being caught using various destructive fishing gears. That is why; the hilsa fishery in Bangladesh has been suffered by a combination of factors *viz.* serious recruitment over-fishing (indiscriminate harvest of gravid fishes) and growth over-fishing (indiscriminate catching of jatka). In these circumstances, considering the importance of hilsa in nutrition, employment and economy, the Hilsa Fishery Management Action Plan (HFMAP) was prepared for the development, management and conservation of hilsa incorporating the objectives of protection of nursery and breeding grounds and banning capture of hilsa indiscriminately. Hilsa is caught and landed throughout the year; the majority of landing (60-70%) is found during the peak breeding season (September-October). In this season, about 60-70% hilsa are found to be sexually mature and ripe. At least 30% of the population appears to be ripening at any time in most areas.

Five sites in the coastal areas of the country have been declared as hilsa sanctuaries under the 'Protection and conservation of fish Act-1950 for the effective conservation of jatka in the major nursery areas and maintenance of fish biodiversity *viz,* 1. From Shatnol of Chandpur district to char Alexander of Laxmipur (1000 km of lower Meghna estuary) 2. Madanpur/Char Illisha to Char Pial in Bhola district (90 km area of Shahbajpur river, a tributary of the Meghna river). 3. Bheduria of Bhola district to Char Rustam of Potuakhali district (nearly 100 km area of Tetulia river). 4. Whole 40 km stretch of Andharmanik river in kalapara Upazila of Potuakhali district and 5. Lower Padma river at Shariatpur district, 20 km stretch of Padma river. It has been identified that the highest number of ripe and running hilsa are being caught indiscriminately during five days before and nine days after the full moon including the full moon day altogether of September-October every year during their peak spawning time and thus their recruitment was being hampered. Hence, fishing ban is required for certain time specification for their successful breeding. In the above context, Government has enacted a new rule under the Protection and Conservation of Fish Act, 1950 banning the hilsa catch during this period for successful spawning. The rule is being implemented by Department of Fisheries (DoF) from the year of 2007 involving different stakeholders and law enforcing agencies including Navy and Coast Guards.

MATERIALS AND METHODS

The hilsa investigation team of the Riverine Station, Chandpur carried out the investigations. Modernized research vessel and speed boat were utilized for sampling and data collection. Major spawning grounds of hilsa and related areas *viz,* Chandpur, Ramgoti, Hatia, Dhalchar, Moulovirchar, Monpura, Kalirchar, Daulatkhan, Barisal, Bhola, Patharghata, Potuakhali, Mohipur, Galachipa, Kalapara were visited for comprehensive study. To determine the size of group, captured hilsa from the above mentioned area were calculated by measuring scale. Sex was determined by external observation, gentle stripping at their belly along the ventral scute line from anterior to anal direction with the fore and first finger together was applied. While at stripping along the belly, white milky or creamy liquid for the male and eggs with blood strain or food particle for the female usually came out through the anus. Seeing the milky/creamy liquid (the milt) and the eggs, hilsa was identified as male or female. When such milky liquid or eggs do not come out, and then the fish is either of immature or premature stage. The potbellied, reddish and bigger anus also identified the fishes as female. As such, the percent composition of gravid and oozing hilsa was determined (Rahman et al., 2013). The spent fishes were identified by observing their very lean and thin and elongated body and health condition and shrunken belly. While stripping at their belly, isolated and distorted eggs came out with watery liquid or with or without blood strain. The hilsa are the gonochoristic (Blaber et al., 2001) and single shedding fishes (Haldar, 2004). After shedding usually they do not die, locally called Pite (spent) fishes and are caught with other hilsa. The number and percent of spent fishes was determined by observing the catches of the commercial fishermen in and around the spawning grounds immediate before and after the 15 days ban period. The amount of fertilized egg production in the spawning grounds due to fishing ban as an indicator of spawning success was estimated using the following formula.

Total No. of Hilsa excluded due to fishing ban (TN) = No. of fishing boat × Haul/day × Fish

Caught/Haul × No. of days -- (1)

Total fertilized eggs (Kg) = $\dfrac{TN \times FF \times SF \times EF}{1000}$-- (2)

Where, TN = Total No. of Hilsa excluded due to fishing ban;
FF = % of female fishes in the study area;
SF = % of spent Fish, and
EF = Average egg (g) per fish.

Experimental egg/fry collection was done by a 'savar net' (shrimp PL collecting net) prepared by fine meshed glass nylon in the spawning grounds and adjacent areas to observe abundance and distribution of jatka. Finally, by the availability of immature and oozing hilsa statement about the previously identified spawning grounds were made.

RESULTS AND DISCUSSION

Size, sex and percent composition of berried hilsa

During the river cruise gradually larger sized hilsa were found from Chandpur to the downstream. In the upper region, most of the hilsa found were below 35 cm, whereas, more than 85% hilsa were above 35 cm in the downstream areas. In and around the spawning grounds among all the captured hilsa, male: female ratio was 1:1.86 and percent composition was 35% and 65% respectively. Haldar (2004) also found male-female ratio almost 1:2 during the study period. Although, there are conflicting views about the sex ratio of hilsa in earlier studies, (Islam et al., 1987) found no significant difference of male-female ratio at four important landing centers *viz* Chittagong (1:1.04); Chandpur (1:1.08); Khepupara (1:0.8) and except Cox's Bazar (1:1.8). Similar observations were also made by (Shafi et al., 1976) in respect of Padma and Meghna river hilsa. Quereshi (1968) observed the sex ratio was 1:1 during the monsoon, but female was dominated in October. Blaber et al., (2001) indicated that here is a bias in sex ratio and the male are more abundant among the smaller fishes. Rahman et al., 2013 found the majority of fishes over 32 cm are females and almost all over 38 cm and males are predominant between 10 and 25 cm length group, the present findings support these views.

From present study it could be seen that, in the spawning ground areas (Monpura and Hatia) more than 85% fishes are above 35.0 cm sizes, which are breeder group of hilsa. Rahman et al., (2013) also found more than 95% hilsa in these areas were above 32 cm in size. Present studies also reconfirm these areas was major spawning grounds of hilsa.

The percent composition of oozing/berried hilsa

In the spawning grounds, hilsa fishes were found with higher maturity stages/berried (maturity stages IV, V and VI) than the other adjacent areas and percent of the mature hilsa were found higher with the higher length group of fishes due to fishing ban in the spawning grounds (Table 1). Similarly, plenty of fries and juveniles of other fishes were also found in and around the spawning ground areas indicating a positive impact of fishing ban on their successful reproduction. Rahman et al., (2013) found higher maturity stages/berried (maturity stages V and VI) in the breeding areas and percent of the mature hilsa were found higher with the higher length group of fishes. The present study agreed with the findings of Rahman et al., (2013).

Table 1. Maturity stages of hilsa at different length group

Length group	Percent of hilsa at maturity stages (Ms)					
	Ms I	Ms II	Ms III	Ms IV	Ms V	Ms VI
18-24	0.00	0.00	60.00	20.00	20.00	0.00
25-31	0.00	7.78	19.85	37.21	29.20	5.96
32-38	0.00	0.00	7.38	31.84	32.68	28.10
39-45	0.00	6.90	0.00	20.75	65.25	7.10
46-52	0.00	0.00	0.00	17.65	82.35	0.00

Percent composition of spent hilsa and spawning success

Catch composition obtained from the major spawning grounds revealed that more than 90% captured hilsa weighing around 900 gm were gravid. In the year 2010, 2011, 2012, 2013, 2014 and 2015 about 33.69%, 36.27%, 35.79%, 41.02%, 38.79% and 36.6%, respectively spent hilsa was observed in the fish landing centers and this data was compared to the data of GEF-BFRI studies (Haldar, 2004) and was found about 67.38, 72.54 , 71.58, 82.04 and 77.58 times higher than that of 2002. Rahman et al.,(2013) found more than 95% captured hilsa weighing around 1.0 kg were gravid and about 5% spent hilsa. The comparative study showed that 15 days fishing ban during spawning seasons might have significant role in the successful reproduction of hilsa.

Figure 1. Pictorial view of spent hilsa

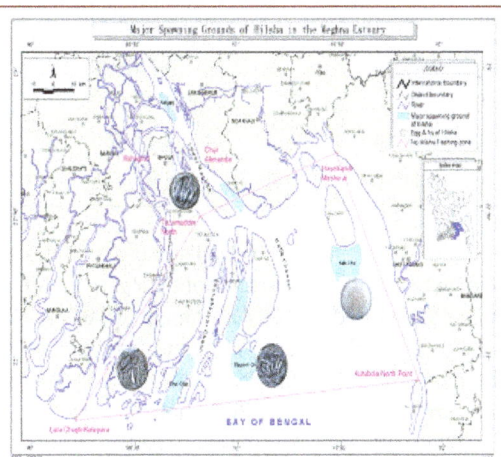

Figure 2. Major spawning grounds of hilsa

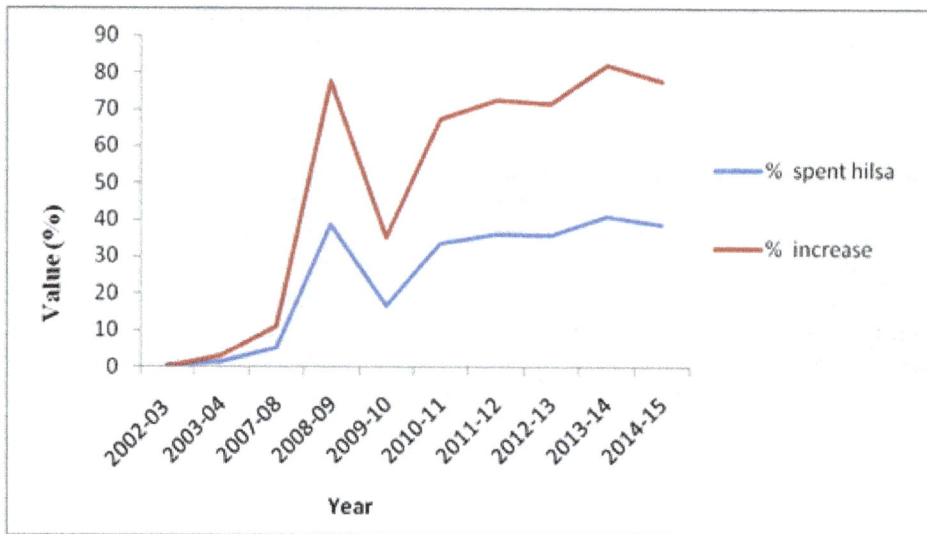

Figure 3. Year wise percentage of spent hilsa. Rahman et al., (2008 and 2013)

Egg/fry production during fishing ban

Present study showed that in the year 2010, 2011, 2012, 2013, 2014 and 2015 hilsa eggs were produced about 336199 Kg, 385500 Kg, 380400 Kg, 447100 Kg, 417765 Kg and 494365 Kg, respectively due to imposing of 15 days fishing ban in spawning season (Table 2). Rahman et al., (2013) found about 46,800 Kg eggs were produced due to imposing of 10 days fishing ban in spawning season in 2007. So, the present study showed the increasing trend. Rahman et al., (2013) found during experimental spawn and juveniles collection, approximately 5-25 days old fries were found in all the surveyed areas in and around the spawning grounds. Present study also showed the similar trend of result (Table 3).

Table 2. Eggs and jatka Production of hilsa in the major spawning grounds. Rahman et al., (2008 and 2013)

Year	Egg production (kg)	No of fries (crore) (50% hatching)	Production of jatka (crore) (10% survival)
2008	3,92,620	2,45,385	24,538
2009	1,70,420	1,06,512	10,651
2010	3,36,199	2,10,124	21,012
2011	3,85,500	2,40,937	24,094
2012	3,80,400	2,37,750	23,775
2013	4,47,100	2,79,437	27,943
2014	4,17,765	2,61,103	26,110
2015	4,94,365	3,08,978	30,897

Due to impose of 15 days fishing ban in the spawning grounds, comparatively higher percentage of gravid hilsa were found which were not available in the similar quantity and condition in the previous years, *i.e.* during fishing ban period (Table 4). Rahman et al., (2013) found also the same result.

Haldar (2004) found complete fishing ban has a strong positive impact on jatka abundance. Similarly, Rahman et al., (2013) found jatka abundance increase 570% by 10 days fishing ban. Present study found that the number of jatka fries (50% hatching) in the year 2010, 2011, 2012, 2013, 2014 and 2015 were 2,10,124, 2,40,937, 237750, 2,79,437, 2,61,103 and 3,08,978 crore (Table 2), respectively. Rahman et al., (2013) found huge number of fries and juveniles of hilsa by 10 days fishing ban.

Table 3. Size, weight and age of captured fries and jatka from the major spawning grounds

Location	Lowest size (cm)	Highest size (cm)	Avera ge size (cm)	Minimum weight (g)	Maximu m weight (g)	Average weight (g)	No. of jatka/haul	Approx. age (day)
Ramneaz	1.57	2.55	1.85	0.01	0.14	0.07	22	10-15
Sakuchia	0.95	3.40	2.12	0.04	0.46	0.13	32	5-10
Janata (ht*)	1.85	3.60	2.70	0.07	0.55	0.22	47	15-25
Janata (lt**)	1.80	3.50	2.65	0.05	0.54	0.19	31	15-20
Hazirhat(ht*)	1.67	3.80	2.62	0.04	0.57	0.15	58	15-20
Hazirhat(lt*)	1.65	3.60	2.60	0.04	0.45	0.16	55	15-20
Dhalchar	2.20	2.9	2.60	0.13	0.30	0.15	46	20-25
Noyar hat	1.7	4.2	3	0.20	0.66	0.28	60	15-20
Horina	1.5	2.8	2.1	0.05	0.50	0.20	24	15-20
Dhulkhola	1.66	3.6	2.8	0.06	0.55	0.18	38	15-20

ht* – high tide, lt** – low tide

Table 4. CPUE of jatka and management strategy in the major spawning grounds (Rahman et al., 2008 and 2013)

Year	Caught jatka/100 m net/hour (Kg)	Comment
2002	0.94	Sanctuary
2005	0.94	Sanctuary
2006	0.61	Sanctuary
2007	0.72	Sanctuary
2008	1.89	Sanctuary + 10 days Ban
2009	2.31	Sanctuary + 10 days Ban
2010	2.44	Sanctuary + 10 days Ban
2011	2.72	Sanctuary + 11 days Ban
2012	2.74	Do
2013	2.77	Do
2014	3.04	Do
2015	3.07	Sanctuary+ 15 days Ban

Most of the fishers refrained from fishing during the ban period in the ban areas. Jatka fry, fries of other fishes were also found plentiful in the spawning grounds and adjacent areas and thus it is assumed that fishing ban have some positive impact on the successful breeding of other fishes. Rahman et al., (2008 and 2013) also found same observations.

CONCLUSION

To sustain as well as to increase hilsa production several management measures have been undertaken by the Ministry of Fisheries and Livestock based on the previous research findings and recommendations. Among the different attempts, conservation of jatka through declaring five fish sanctuaries in the major nursery and spawning grounds of river system and protection of berried hilsa catches for 15 days during the peak breeding season are the most important initiatives. The fishing ban was found effective for successful breeding of hilsa. The impact of 15 days fishing ban on breeding success of other fishes needs to be assessed. The fishing ban should be continued for sustainable reproduction of hilsa and increase of jatka as well as hilsa production.

REFERENCES

1.   Al-Baz, AF and DJ Grove, 1995. Population biology of Sbour *Tenualosa ilisha* (Hamilton-Buchanan) in Kuwait. Asian Fisheries Science, 8: 239-254.

2.   Blaber, S. J. M. 2000. Tropical estuarine fishes. Ecology, exploitation and conservation, Blackwell, Oxford, 372 pp.

3.   Blaber SJM, DA Milton, DT Brewer and JP Salini, 2001. The shads (genus *Tenualosa*) of tropical Asia: An overview of their biology, status and fisheries. In: Proc., international Terubok. Conf. Sarawark Development Institute (SDI), Kuching, Sarawark, P. 9-17.

4.   Fisher W and G Bianchi (Eds.), 1984. FAO species identification sheet for fishes. Vol.1, *Tenualosa ilisha* (Clupeidae). Food and Agricultural Organization of the United Nations, Rome, Italy.

5.   Haldar GC, M Rahman and AKY Haroon, 1992. Hilsa, *Tenualosa ilisha* (Hamilton) Fishery of Feni River with reference to the impacts of the flood control structures. Journal of Zoology, 7: 51-56.

6.   Haldar GC and MA Rahman, 1998. Ecology of Hilsa, *Tenualosa ilisha* (Hamilton). In: Proceedings of BFRI/ACIAR/CSIRO Workshop on Hilsa Fisheries Research in Bangladesh, held on 3-4 March 1998 at Bangladesh Agricultural Research Council, Dhaka, Bangladesh (edited by M.A. Mazid and S.J.M. Blaber). BFRI proc. No. -6: 86pp.

7.   Haldar GC, 2004. Present status of the hilsa fisheries in Bangladesh: Report on hilsa management and conservation studies conducted under the ARDMCS, GEF component, FFP, 70 pp.

8.   Haldar GC, 2008. Hilsa Fisheries Conservation, Development and Management Technique. 40pp.

9.   Hamilton F, 1822. An account of the fishes found in the river Ganges and its branches. Edinburgh & London. An account of the fishes found in the river Ganges and its branches. : I-VII + 1-405, Pls. 1-39.

10. Islam MS, Haq QM, Hossain M, Azad SA and NN Das, 1987. Hilsa fishery in Bangladesh in 1985-1986. Hilsa Investigations in Bangladesh. FAO-UNDP Proj. Mar.Fish.Resour. Manage. Bay of Bengal, Colombo, Sri Lanka. BOBP/Rep/36, 81-95 Rahman et al.

11. Rahman MA, BMS Rahman, SJ Hasan, Flura, T Ahmed and MI Haidar, 2013. Impact of Eleven Days Fishing Ban in the Major Spawning Grounds of Hilsa (*Tenualosa ilisha)* (Hamilton) on its Breeding Success. Bangladesh Research Publications Journal, 9: 116-122.

12. Quereshi MR, 1968. Hilsa Fishery in East Pakistan. Pakistan Journal of Science and Industry Research. 11: 95-103pp

13. Rahman MA, MA Alam, KKU Ahmed, T Ahmed and GC Haldar, 2008. Assessment of Impact of Ten days Fishing Ban in the Major Spawning Grounds of Hilsa (*Tenualosa ilisha)* Fisher and Bianchi, 1984). Bangladesh Journal of Fisheries Research, 13: 27-33

14. Raja BTA, 1985. A review of the biology and fisheries of *Hilsa ilisha* in the Bay of Bengal. Bay of Bengal Programme, BOBP/WP/37. Marine Fishery Resource Management in the Bay of Bengal, 66 pp

15. Shafi M, MMA Quddus and N Islam, 1976. Maturation, spawning, sex-ratio and fecundity of *Hilsa ilisha* (Hamilton-Buchanan) of the river Padma. Proc. First Bangladesh Science Conference Abstracts: B-45.

16. Whitehead PJP, 1985. FAO Species Catalogue 7, Clupeids fishes of the World. Part I Chirocentridae, Clupeidae and Pristigasteridae. FAO Fisheries Synopsis 125, Volume 7, Part 1, 303 pp.

8

HISTO-ARCHITECTURAL CHANGES OF INTESTINAL MORPHOLOGY IN ZEBRA FISH (*Danio rerio*) EXPOSED TO SUMITHION

Sk. Istiaque Ahmed[1,2], Mirja Kaizer Ahmmed[2], Subrata Kumar Ghosh[2], Md. Moudud Islam[2] and Md. Shahjahan[1*]

[1]Department of Fisheries Management, Faculty of Fisheries, Bangladesh Agricultural University, Mymensingh-2202, Bangladesh; [2]Faculty of Fisheries, Chittagong Veterinary and Animal Sciences University, Khulshi, Chittagong- 4225, Bangladesh

*Corresponding author: Dr. Md. Shahjahan; E-mail: mdshahjahan@bau.edu.bd

---

ARTICLE INFO	ABSTRACT

Key words

Sumithion
Intestine
Zebra fish
Histology
Pesticide

Organophosphorous pesticide sumithion, the O, O Dimethyl O- (3-methyl-4-nitrophenyl), is a widely used pesticide in agricultural land and in aquaculture to control some harmful pests. A study was conducted in laboratory condition with aquaria (36 inch × 10 inch × 12 inch) to evaluate the effects of sumithion on histo-architecture of intestine in zebrafish (*Danio rerio*). The experiment was carried out with three treatments (T_1: 0.5 ppm, T_2: 1.0 ppm, T_3: 2.0 ppm) and a control (T_0: 0ppm), each having three replications. Zebra fishes (*Danio rerio*) (4±1cm and 0.9±0.2g) were stocked for the experiment and sacrificed after 7 days of exposure of sumithion. During the study period, the temperature was almost constant (21-22°C) but dissolved oxygen, pH and total alkalinity values were tended to decrease with the increase in concentrations of test chemicals. The histo-architectural changes in intestine suggested that the intestinal epithelial cells, lumen, villi and intestinal folding were varied significantly ($P<0.05$) in treatment groups rather than the control groups (T_0). Disappeared mucosa (DM) along with abnormal lumen (AL) were found in case of T_1, while destructed intestinal villi (DV), sloughing of superficial epidermal cells (SEC) and uneven intestinal folding (UF) were found in T_2 and T_3. The obtained result supports the toxic potentiality of sumithion. Therefore, the use of sumithion must be evaluated carefully in agriculture and aquaculture.

INTRODUCTION

Pesticides have been considered as a major threat of gradual degradation for the aquatic ecosystem (Salam et al., 2015; Sharmin et al., 2015). Thirty nine insecticides, under four major groups *viz.* organochlorine, organophosphate, carbamate and pyrethroid, are being used in agricultural and public health sector (Satter, 1985) of which over 98% of sprayed insecticides and 95% of herbicides have impacts not only upon their target species but also on non-target species, air, water, bottom sediments and food (Miller and Miller 2004; Rahman et al., 2012).

Sumithion, an organophosphate insecticide is considered somewhat toxic to fish (Thomson, 1989) but no clear evidence on the intestinal morphological alteration was found so far. Toxicology studies on different tissues (Thophon et al., 2003) can provide information about tissues injuries and damages of organs resulting in morphological dysfunction.

Being one of the most common, small and robust vertebrate model organisms, zebrafish got the highest priority to conduct this experiment (Shahjahan et al., 2013). Generally, every species possess an immense power of biotic potential in favorable condition by increasing the population to the explosive level. But, in adverse condition, different organs such as intestine, gonad etc. may be affected. Therefore, this study was carried out to determine the effect of sumithion on intestine of zebrafish.

MATERIALS AND METHODS

Proper and absolute analysis of impacts of pesticides and their contamination under field conditions is extremely difficult, but the need for field tests in this aspect can be reduced to a greater extent by the implementation of controlled laboratory tests. That is why, the present study was carried out from January to April, 2014 at the wet laboratory of the Faculty of Fisheries, Bangladesh Agricultural University (BAU), Mymensingh, Bangladesh.

Species selection

The Zebrafishes (*Danio rerio*) with a length of 4±1 cm and weight of 0.9±0.2g, were selected on the basis of the health condition from different ponds adjacent to academic building of Fisheries Faculty, BAU. The selected fishes were acclimatized in aquaria at 22 ± 0.5°C under a controlled natural photo-regimen (14/10 h, light/dark) condition for a period of 21 days before the experiments. During acclimatization process, the fish were fed twice a day with commercial grower feed (CP Bangladesh Co., Ltd.).

Experimental procedure for acute and behavioral toxicity

According to the standard method, a static acute toxicity bioassay was performed to determine the 24, 48, 72, and 96 h lethal concentration values (LC50) of sumithion for Zebra fish. Seven different concentrations (4, 5, 6, 7, 8, 9, and 10 ppm) of sumithion with three replicates were used in the test series. Control units with three replicates were also prepared. Exceeding aeration was applied to the aquarium for 2 h in order to obtain a homogeneous concentration of the toxic compound, and then 10 fish were transferred into each aquarium. Mortality was assessed at 24, 48, 72, and 96 h after the start and dead fishes were removed immediately. Several behavioral changes, such as reduced activity, equilibrium imbalance, abnormal swimming and motion inactivity of the fishes were observed during the exposure period.

Experimental design

Twelve aquaria (36 inch × 10 inch × 12 inches) were collected, cleaned, washed and sun-dried properly prior to set in the wet laboratory. The experiment was conducted with three treatments (T_1: 0.5 ppm, T_2: 1.0 ppm, T_3: 2.0 ppm) and a control (T_0: 0ppm), each having three replications. Ten fish were stocked in each aquarium containing 20L of tap water. Fish were sacrificed after the desired exposure of sumithion (7 days) to observe the effects on intestine. The fish sample was collected and fixed in 10% formalin for the use of histological analysis. The application of pesticide at desired concentration was reapplied at every 24 h with a regular exchange of water.

Feeding frequency

The feeding frequency was 2 times per day (9.00am and 9.00pm) at a rate of 70% of their body weight throughout the experimental period. Before introducing feed for the next feeding, previous uneaten feeds and feces in aquarium were removed by siphoning using a plastic pipe.

Monitoring of water quality parameters

During the experimental period, the water quality parameters such as temperature with a thermometer, pH with a pH meter, dissolved oxygen with a DO meter and total alkalinity with hachkit were recorded.

Histological study of intestine

For histological analysis of intestine, three fish species were selected from each treatment. Then cephalic and caudal portion of the selected fishes were cut off and the intestinal section was preserved at 10% formalin. The preserved samples were taken out from vials and put into cassettes separately. Then dehydration process was carried out manually followed by clearing, infiltration, embedding, sectioning, staining and mounting. Finally, intestinal sections were observed under microscope and photographs were taken at 10x magnification.

Statistical analysis

Values were expressed as means ± standard deviation (SD). Data were analyzed by one-way analysis of variance (ANOVA) followed by Tukey's post hoc test to assess statistically significant differences among the control and different treated values. Statistical significance was set at $P < 0.05$. Statistical analyses were performed using PASW Statistics 18.0 software (IBM SPSS Statistics, IBM, Chicago, USA).

RESULTS AND DISCUSSIONS

Acute and behavioral toxicity

The mortality patterns of the test species exposed to different doses of sumithion are presented in Table 1. No mortality was observed in control treatment, whereas the mortality percentage increased as the concentration of sumithion increased. The LC50 value of zebrafish for sumithion during the 96 h of exposure was 7.89 ppm (≈8). A series of abnormal behavior such as restlessness, sudden quick movement, rolling movements, swimming on the back (at higher doses) etc. was observed during the continuation of experiment. Due to the application of higher doses, the affected fish became extremely weak and ultimately died. No such abnormalities were observed in terms of behavior in the control group.

In the present study, the LC50 value (7.89 ppm) recorded for *D. rerio* is less than the values (9.14 ppm for *Ptychocheiilus lucius*, 11700 µg/L for black bullhead, 11.8 ppm for *Heteropneustes fossilis*, 15.3 ppm for *Gila elegance* and 17.0 ppm for *Ictalurus furcatus*) determined by Durkin (2008) and Faria *et al.* (2010) for different fish species. In contrast to the above-mentioned values, Pathiratne and George (1998) reported a lower 96 h LC50 value (2.2 ppm) for *Oreochromis niloticus*. Newhart (2006) tabulated the LC50 values of malathion for different species of fish which ranges from 0.06 to 7620 µg/L. Malathion was found to be highly toxic to fry of *Labeo rohita* (LC50 value 9 µg/L), Patil and David, (2008); *Opheocephalus punctatus* (LC50 16 µg/L), Pugazhvendan *et al.*, (2009); walleye (LC50 64 ppb), brown trout (LC50 101 ppb) and cutthroat trout (LC50 280 ppb) and moderately toxic to minnows (LC50 8.6 ppm) and murrels (LC50 5.93 ppm) as summarized by Durkin (2008). The difference in the potentiality of pesticides toxicity may be attributed mainly to the susceptibility of the test animals and several factors like pH and hardness of water. The observation of a series of abnormal behaviors such as restlessness, loss of equilibrium, increased opercular activities, surface to bottom movement, sudden quick movement, resting at the bottom, etc. were similar to the observations of Haque *et al.* (1993) and Lovely (1998). Some aspects that are contrary to the findings of Kabir and Begum (1978) and Lovely (1998) such as swelling in the abdominal region and gas-filled stomach were not observed.

Effects of sumithion on water quality parameters

Recording of water quality parameters (Temperature, Dissolved oxygen, pH and Total Alkalinity) was a regular task during the exposure period of sumithion at various concentrations as well as with control. During the study period of 7 days the values of Temperature were almost constant regardless of the application of treatments. Dissolved oxygen, pH and total alkalinity values were tended to decrease with the increase in concentrations of test chemicals (sumithion) (Table 2), but the values were not significantly varied (P> 0.05).

Table 1. Mortality percentages of the selected fish samples exposed to different concentrations of sumithion at different time intervals.

SL No.	Concentration (mg/L)	Initial No. of fish	Count of dead fish after				% of mortality
			24 h	48 h	72 h	96h	
1	Control	10	-	-	-	-	00
2	4.0	10	-	-	-	1	10
3	5.0	10	-	-	-	2	20
4	6.0	10	-	-	2	3	30
5	7.0	10	1	2	3	4	40
6	8.0	10	-	2	4	5	50
7	9.0	10	1	3	7	9	90
8	10.0	10	2	7	9	10	100

Table 2. Water quality parameters (Means ± SD) during the study period.

Parameteres	Treatments			
	T_0 (0 ppm)	T_1 (0.5 ppm)	T_2 (1.0 ppm)	T_3 (2.0 ppm)
Temperature (°C)	21.6± 0.71	21.48 ± 0.34	21.34 ± 0.15	22.20 ± 0.40
pH	7.21 ± 0.07	6.81 ± 0.07	6.58 ± 0.11	6.40 ± 0.13
Dissolved oxygen (mg/L)	4.62 ± 0.12	3.41 ± 0.15	3.18 ± 0.19	2.93 ± 0.12
Total Alkalinity (mg/L)	175.71 ± 1.38	163.57 ± 1.98	159.28 ± 6.36	149.28 ± 6.89

There have been profound significances on environmental parameters in affecting the toxicity of different pesticides. Considering the significance of temperature on different factors like enzyme activity, metabolic rate, oxygen uptake etc, it has been studied more widely than any other environmental parameters. Generally, toxicity is more or less proportionate to high temperature. Macek et al. (1969) studied the effects of 10 pesticides to the rainbow trout and 11 pesticides to the bluegills at different temperatures and found that the toxicity increased with increasing temperature. P^H has also been found significant in influencing various physicochemical properties of pesticides like hydrolysis, volatilization and in balancing the dissociated and undissociated forms (Weber, 1972). The toxicity of organophosphate (OP) compounds is not influenced by pH very commonly except in few cases. The toxic effects of 2, 4-D were reduced when the pH was raised by the addition of sodium chloride (Holcombe et al., 1980). Davies (1975) attempted to formulate the criteria for minimum dissolved oxygen requirement of fish. His approach was on examining the threshold levels of dissolved oxygen that cause changes in some physiological lesions. Channel Catfish exposed for 72h to an oxygen content of 1.5 ppm showed anomalies in gill, liver, kidney and spleen (Scott and Rogers, 1980). In the present investigation the variation of the different water quality parameters (Temperature, Dissolved oxygen, pH and Alkalinity) that were monitored during the exposure period within various concentrations of pesticides as well as with the control were not significant. The limited variations in these parameters among different treatments may be due to regular renewal of water and pesticide at every 24 hours.

Effects of sumithion on histo-architecture of intestine

In this experiment, the histo-architecture of zebrafish intestine was observed through histology. The intestinal epithelial cells, lumen, villi and intestinal folding were almost regular in control groups (T_0), while in treatments various abnormalities were found. In T_1, the intestinal mucosa was almost disappeared (DM) along with abnormal lumen (AL). On the other hand, destructed intestinal villi (DV), sloughing of superficial epidermal cells (SEC) and uneven intestinal folding (UF) were found in T_2 and T_3 (Figure 1). The result also indicated that the Zebra fish intestine was damaged by 20%, 50% and 70% in T_1, T_2 and T_3, respectively (Figure 2) and varied significantly ($P<0.05$) among treatments.

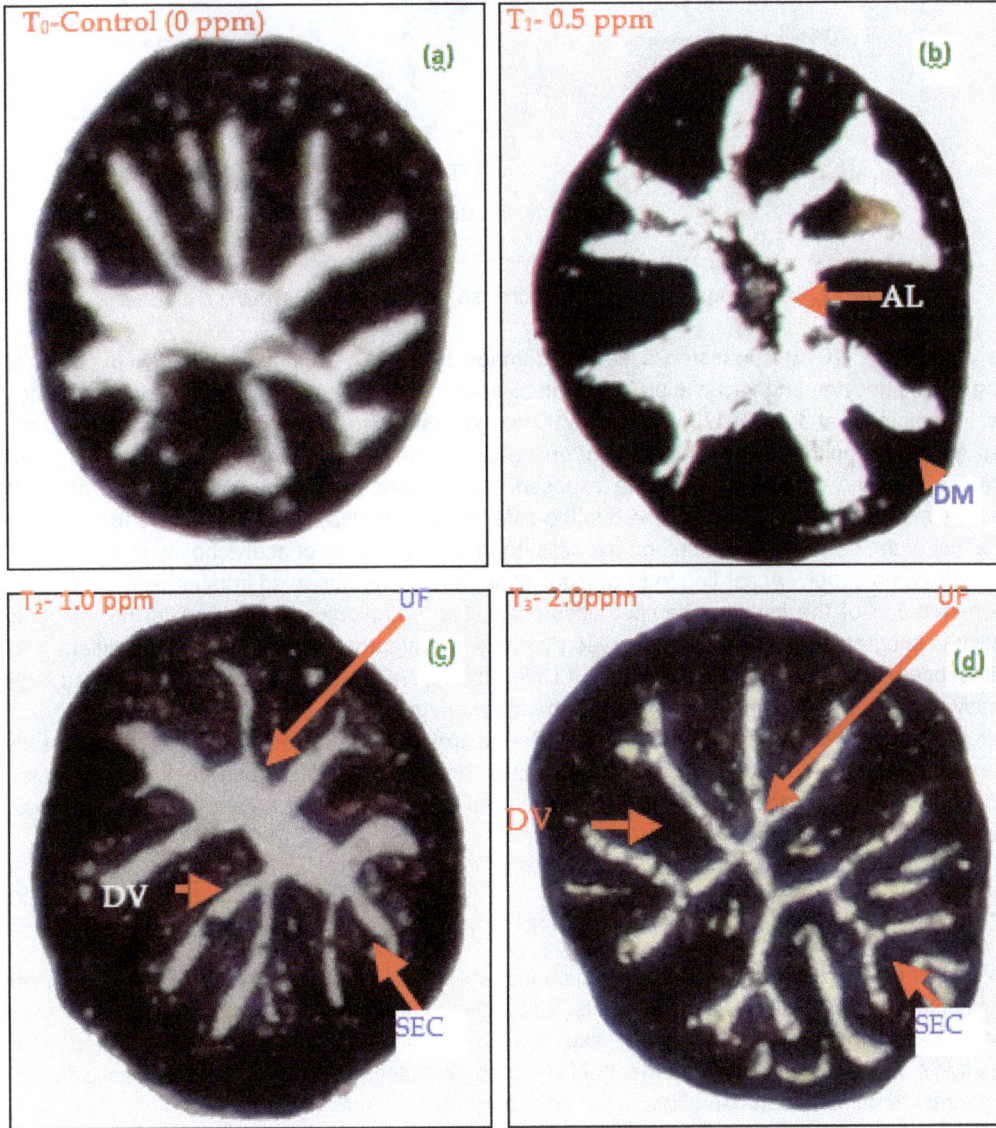

Figure 1. Histo-architectural changes in intestine exposed to sumithion; (a) Control (T_0), (b) 0.5 ppm (T_1), (c) 1.0 ppm (T_2) and (d) 2.0 ppm (T_3). Arrowheads are indicating Abnormal Lumen (AL), Disappeared Mucosa (DM), Destructed Villi (DV), Uneven intestinal Folding (UV) and Sloughing of Epithelial Cells (SEC).

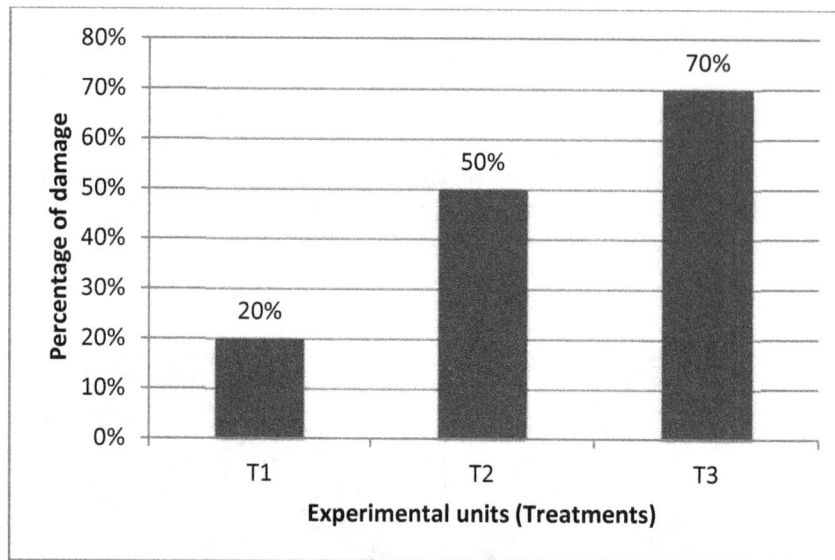

Figure 2. Percentage of damage in Zebra fish intestine due to sumithion exposure

Das and Gupta (2013) suggested a severe damage in intestinal mucosa, destructed epithelial walls, infiltration of lymphocytes and vacuolation in the intestine of Indian Flying Barb, *Esomus danricus* when it was exposed to malathion at 1.79, 0.179 and 0.00179 µg/l concentration for 28 days. Velmurugan *et al.* (2007) observed atrophy of epithelial cells, necrosis of epithelial cells and infiltration of lymphocytes into the lamina propia in the intestine of *Cirrhinus mrigala* exposed to fenvalerate. According to Desai *el al.* (1984), the degenerative and necrotic changes observed in the different intestinal layers of the studied fish may be due to a direct effect of the detected pesticides on the cells, to an accumulation of acetylcholine in the tissues or to a reduction in oxygen supply. According to Bhatnagar *et al.* (2007), the observed irritation and destruction of the mucosa membrane of the intestine hamper absorption. The pathological alterations in the intestine of the studied fish is in agreement with those observed by many investigators about the effects of different toxicants on fish intestine (Hanna *et al.*, 2005; Cengiz and Unlu, 2006). Walsh and Ribelin (1975) reported hyperemia, degenerative changes in the tips of villi, loss of structural integrity of mucosal folds, hypertrophy vacuolation and necrosis in the intestine of *Cyprinus carpio* exposed to the pesticide atrazine. The present study, thus, concludes that although organophosphate pesticides are found to be less toxic to mammals than organochlorines, yet, very low doses of such pesticides (sumithion) can cause apparent damage in the intestine of fish.

CONCLUSION

The current research activity was conducted on the basis of an aim to evaluate the effects of sumithion on histo-architecture of intestine in zebrafish. It is concluded that sumithion seems to be somewhat toxic to zebrafish. The data obtained in the present experimental protocol revealed significance that sumithion has adverse effects on the intestinal arrangement of fish. In addition, the indiscriminate use of pesticide in the field may be a threat to human, fauna and flora of the environment.

CONFLICT OF INTEREST

The authors declare that they have no conflict of interests. The authors alone are responsible for the content and writing of the paper.

ACKNOWLEDGEMENTS

This work was supported by a grant of Impact of Aquaculture Drugs and Chemicals on Aquatic Ecology and Productivity Project (IADCAEPP) provided by Bangladesh Fisheries Research Institute (BFRI), Mymensingh-2201, Bangladesh.

REFERENCES

1. Bhatnagar C, Bhatnagar M and Regar B, 2007. Fluoride-induced histopathological changes in gill, kidney, and intestine of fresh water teleost, *Labeo rohita*. Research Report Fluoride, 40: 55-61.
2. Cengiz EI and Unlu E, 2006. Sublethal effects of commercial deltamethrin on the structure of the gill, liver and gut tissues of mosquitofish, *Gambusia affinis*: a microscopic study. Environmental Toxicology and Pharmacology, 21: 246-253.
3. Das S and Gupta A, 2013. Histopathological changes in the intestine of Indian Flying Barb (*Esomus Danricus*) exposed to organophosphate pesticide, Malathion (Ec 50). Global Journal of Biology, Agriculture and Health Sciences, 2: 90-93.
4. Davies JS, 1975. Minimal dissolved oxygen requirement of aquatic life with emphasis on Canadian species: a review. Journal of Fisheries Research Board of Canada, 32: 2295-2332.
5. Desai AK, Joshi VM and Ambadkar PM, 1984. Histological observations on the liver of *Tilapia mossambica* after exposure to motiocrotophos, an organophosphorus insecticide. Toxicolology Letters, 21: 325-331.
6. Durkin PR, 2008. Malathion; Human Health and ecological risk assessment. Final report submitted to Paul Mistretta, PCR, USDA/Forest Service, Suthern region, Atlanta, Georgia, pp. 325.
7. Faria IR, Palumbo AJ, Fojut TL and Tjeerdema RS, 2010. Water Quality Criteria Report for Malathion. Phase III: Application of the Pesticide Water Quality Criteria Methodology. University of California, DAVIS, 7, pp. 64.
8. Hanna M, Shaheed I and Elias N, 2005. A contribution on chromium and lead toxicity in cultured *Oreochromis niloticus*. Egyptian Journal of Aquatic Biology and Fish, 9: 177–209.
9. Haque MM, Mirza MJA and Miah MS, 1993. Toxicity of Diazinon and Sumithion to *Puntius gonionotus*. Bangladesh. Journal of Training and Development, 6: 19-26.
10. Holcombe GW, Flandt JT and Phipps GL, 1980. Effect of pH increases and sodium chloride addition on the acute toxicity of 2-4 dichlorophenol to the fathead minnow. Water Research, 14: 1073.
11. Kabir SMH and Begum R, 1978. Toxicity of three organophosphorus insecticides to Singhi fish *Heteropneustes fossilis* (Bloch). Dhaka University Studies, Part B, 26: 115-122.
12. Lovely F, 1998. Toxicity of Three Commonly Used Organo-Phosphorus Insecticides to Thai Sharpunti (*Barbodes gonionotus*) and African Catfish (*Clarias gariepinus*) Fry. M.Sc. Thesis, Department of Fisheries Biology and Genetics, Bangladesh Agricultural University, Bangladesh.
13. Mckim JM, Benoit DA, Biesinger KK, Brungs WA and Siefert RE, 1975. Effects of pollution on fresh water fish. Journal of the Water Pollution Control Federation, 47: 1711-1764.
14. Miller KV and Miller JH, 2004. Forestry herbicide influences on biodiversity and wildlife habitat in southern forests. Wildlife Society Bulletin, 32: 1049–1060.
15. Newhart KL, 2006. Environmental Fate of Malathion. California Environmental protection Agency, USA, pp. 20.
16. Pathiratne A and George SG, 1998. Toxicity of malathion to Nile tilapia, *Oreochromis niloticus* and modulation by other environmental contaminants. Aquatic Toxicology, 43: 261-271.
17. Patil VK and David M, 2008. Behaviour and respiratory dysfunction as an index of malathion toxicity in the freshwater fish *Labeo rohita* (Hamilton). Turkish Journal of Fisheries and Aquatic Sciences, 8: 233-237.
18. Pugazhvendan SR, Narendiran NJ, Kumaran RG, Kumaran S and Alagappan KM, 2009. Effect of Malathion Toxicity in the Freshwater Fish *Ophiocephalus punctatus*-A Histological and Histochemical Study. World Journal of Fish and Marine Sciences, 1: 218-224.
19. Rahman MS, Shahjahan M, Haque MM and Khan S, 2012. Control of euglenophyte bloom and fish production enhancement using duckweed and lime. Iranian Journal of Fisheries Sciences, 11: 358-371.

20. Salam MA, Shahjahan M, Sharmin S, Haque F and Rahman MK, 2015. Effects of sub-lethal doses of an organophosphorous insecticide sumithion on some hematological parameters in common carp, *Cyprinus carpio*. Pakistan Journal of Zoology, 47: 1487-1491.

21. Satter MA, 1985. The use of pesticides in Bangladesh and protection of environment. Proceeding of the SAARC Seminar on Protecting the Environment from Degradation, Dhaka, Bangladesh.

22. Scott AL and Rogers WA, 1980. Histological effects of prolonged sublethal hypoxia of channel catfish, *Lctalurus punctuates*. Journal of Fish Diseases, 3: 305.

23. Shahjahan M, Kitahashi T, Ogawa S and Parhar IS, 2013. Temperature differentially regulates the two kisspeptin system in the brain of zebrafish. General and Comparative Endocrinology, 193: 79-85.

24. Sharmin S, Shahjahan M, Hossain MA, Haque MA and Rashid H, 2015. Histopathological Changes in Liver and Kidney of Common Carp Exposed to Sub-lethal Doses of Malathion. Pakistan Journal of Zoology, 47: 1495-1498.

25. Thomson WT, 1989. Agricultural Chemicals. Book I: Insecticides. Thomson Publications, Fresno, California. pp. 120.

26. Velmurugan B, Selvanayagam M, Cengiz E and Unlu E, 2007. The effects of fenvalerate on different tissues of freshwater fish (*Cirrhinus mrigala* L.). Journal of Environmental Science and Health, 42: 157-163.

27. Walsh AH and Ribellin WE, 1975. The Pathology of Pesticide. In: WR Rebellin and G. Migaki (Editors), Pathology of Fishes. University of Wisconsin press, Madison. pp. 515-537.

28. Weber JB, 1972. Interaction of Organic Pesticides with Particulate Matter in Aquatic Soil system, in Fate of Organic Pesticides in the Aquatic Environment (Advances in Chemistry Series III). Faust, S.D., Eds. American Chemical Society Washington D.C. pp. 55.

9

SURVEY OF SOME ASPECTS OF ARTISANAL FISHERIES OF SABIYEL LAKE, ALIERO, KEBBI STATE, NIGERIA

Ibrahim Shehu Jega[1*], Ibrahim Mohammed Ribah[2], Zakariyya Idris[3], M. Mahfujul Haque[4], Md. Saifullah Bin Aziz[4] and Md. Mehedi Alam[4]

[1]Department Fisheries Management, Faculty of Fisheries, Bangladesh Agricultural University, Mymensingh-2202, Bangladesh; [2]Department of Animal Science, Kebbi State University of Science and Technology, Aliero, Nigeria; [3]Department of Forestry and Fisheries, Kebbi State University of Science and Technology, Aliero, Nigeria; [4]Department of Aquaculture, Faculty of Fisheries, Bangladesh Agricultural University, Mymensingh-2202, Bangladesh

*Corresponding author: Ibrahim Shehu Jega; E-mail: ibrahimshehu77@yahoo.com

ARTICLE INFO

Key words

Artisanal fisheries
Fishing gears Fishing
methods Sabiyel lake
Value addition

ABSTRACT

A survey was conducted to assess some aspects of artisanal fisheries in fishing households of Sabiyel Lake in Aliero local Government Area, Kebbi State, Nigeria between 2014 and 2015. One hundred (100) questionnaires were administered in seven (7) communities surrounding the Lake. Data was analyzed using descriptive statistics. Results revealed that 100% of the sampled households were married. The age of the respondents vary with 43% below the age of 50, 32% above 50 years of age. All of the households sampled were male. Respondents with no formal education accounted for 77%. Majority of the respondents still use traditional gears, methods and crafts. Most respondents (53%) were involved in only selling of fresh product as a means of livelihood. The fish species caught in Sabiyel lake comprises of *Hyperopesus bebe*, *Oreochromis niloticus*, *Sarotherodon galilieus*, *Clarias gariepinus* and *Heterotis niloticus*. Most respondents (42%) sold their fish fresh without processing or preservation. Regarding the role of women, 28% of the respondents mentioned that women participate in the processing of fish. Most of the respondents (87%) stated lack of both modern fishing gears and crafts as the major problems confronting their fishing activities. The study suggested that state government should address the major constraints to fishing in the study area by supplying and subsidizing fishing gears, crafts and adequate processing and preservation equipment.

INTRODUCTION

The Nigeria fisheries industry consist of three (3) broad sub-sectors; the artisanal or small scale fisheries; the industrial (or large scale fisheries) and aquaculture. Out of these three subsectors, artisanal fisheries constitute the most significant sub sector in terms of number of people, employment and contribution to total fish output in the country (Oladimeji et al., 2013). Available records from the Federal Department of Fisheries Statistics reveal that total fish production in Nigeria for 25 years average about 4,08,000 MT per annum. According to Mathieu (2001), artisanal fishing account for more than 80 percent of the total fish production in Nigeria. But in spite of contributing the lion share to domestic fish output in Nigeria, artisanal fisheries remain the most impoverished fisheries sub sector with fishermen generally living at the subsistence level. Several reasons had been offered for the poor standard of living of artisanal fisherman which culminates in various poverty levels in the fishing communities.

Though the riverine communities benefit from species diversity, Bada (2005) noted that Nigeria requires approximately 1.5million MT of fish annually in order to meet its daily protein needs. However, Nigeria has not been able to provide the quantity of fish needed by its citizens and this has led to importation to supplement local production. To stimulate the country to become self-sufficient in fish production over the next four years through a 25% annual fish import cut, an annual baseline fish import figure has been set for 2014 which reduced the allowable quantity of imported fish to 5,00,000 MT (Fishsite, 2014).

Nigeria has been listed among the 25 poorest nations in the world for several years. Yet, the country is endowed with human population of over 160 million people in addition to rich vegetation and abundant water resources; about 2,86,200km2 of waterarea (Shimang, 2005), which apart from capability of supporting a large population of livestock and crop irrigation, as well supports production of enough fish and fish products not only for domestic consumption but also for export (FAO, 2003).Suffice to note also that despite Nigerians abundant fishery resources, the country is still largely a protein deficient nation. It is well documented that Nigerians per capital intake of high quality animal protein is too low (Rahji et al., 2011).

Nigeria is blessed with over 14 million of hectares of reservoirs, lake, ponds and major rivers capable of producing over 9,80,000MT of fish annually (FDF, 2007). Statistical survey have shown that the demand for fish in the country exceed supply and also the domestic production is still very low, considering the increasing human population. The annual fish consumption/demand in Nigeria has been estimated to be over 1.3 millionMT and the total domestic production is just about 4,50,000 MT per annum (Tsadu et al., 2006). With increase in human population in Nigeria, less fish will be available per capita annually (Eyo, 1999). In this regard, several studies have been conducted on the assessment and conservation of various lakes and rivers in Nigeria (Araoye, 2009; Abubakar and Auta, 2012; Ahmed *et al.*, 2014). Despite the rich nature and importance of Sabiyel lake in the provision of animal protein and income to the vast majority of people in the study area, no such research have been conducted.

This research work has the primary goal of identifying the socioeconomic characteristics, type of fishing gears and crafts, dominant fish species caught, processing methods used, size and distribution of fishing communities, role of women in fisheries activities and the social and economic constraints that limit development of the fisheries of Sabiyel lake in order to suggest ways for effective management.

MATERIALS AND METHODS

Study area

The survey was conducted in the fishing communities around Sabiyel lake, Aliero Local Government of Kebbi state. Sabiyel lake is a eutropic, perennial standing fresh water body, located between latitude 13^0 6"-$4^0$15" North and longitude $12^0$27"- 47" East. The lake is at the centre of Sabiyel, Kashinzama, Laga, Tari, Kambaza, Bami Mairuwa, and Kyara villages, 13km away from Aliero town, Aliero Local Government Area, Kebbi state, Nigeria (Figure 1). It covers a length of 9km between the one extreme to other extreme of the lake. It is almost covered by emergent plants, dominantly cattail.

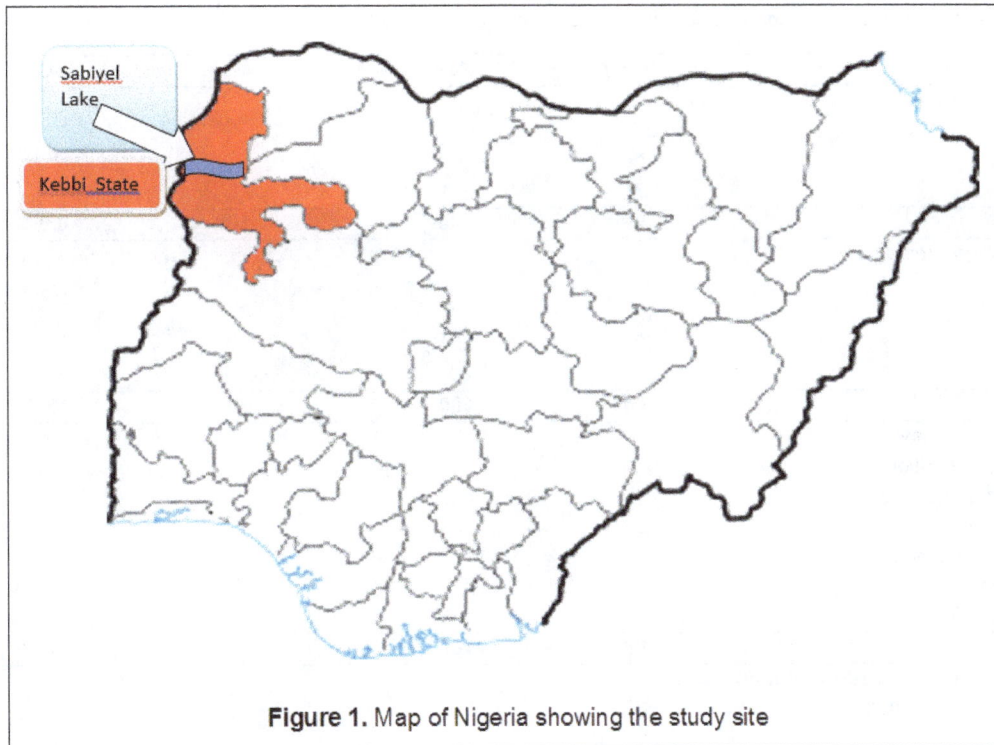

Figure 1. Map of Nigeria showing the study site

Sampling procedure and sample size

There are seven (7) villages surrounding Sabiyel lake. Thus, these villages were purposively selected for this study. Based on the relative number of households in each of the seven (7) villages, a total of 100 respondents were considered. A respondent's involvement in a particular fishing activity was the basis of selection for the interview.The villages covered and numbers of respondents were as follows: Sabiyel (22), Laga (15), Kyara (05), Kashinzama (21), Kambaza (05), Bami mairuwa (15) and Tari (17).

Data collection

The study was based on the primary data obtained from the household heads in the study area with the aid of structured questionnaires. Data were collected on socioeconomic background of fisher folks, information on the type of fish species caught, methods used in processing and preservation of fish, major constraints to fishing activities, fishing rights, roles women play in fisheries and in value addition to fish products.

Data analysis

Data obtained were analyzed with help of descriptive statistics such as percentage, frequency and means using software MS Excel.

RESULTS

Socio economic characteristics of respondents

Socioeconomic characteristics of the respondents are presented in Table 1. The results indicated that 100% of the respondents were married and 75% are between the age ranges of 31-50 years. Majority of the respondents (53%) indicated that their main source of income is the sale of fresh fish products. Most respondents (77%) have attended Qur'anic education with little having formal education. The primary occupation of the respondents is fishing as indicated by 51% respondents.

Results for fishing activities of household members around Sabiyel fishing communities indicated that 86% of the respondents engage in fishing activities for sale and family consumption while7% each engage in fishing either for sale or for family consumption.

Table 1. Distribution of respondents according to socioeconomic characteristics

Characteristics	Frequency	Percentage	Mean	SD
Marital status				
Single	0.0	0.0		
Married	100	100		
Divorced	0.0	0.0		
Total	**100**	**100**	**100**	**0.0**
Age range (years)				
18-30	5	5		
31-40	43	43		
41-50	32	32		
Above 50	20	20		
Total	**100**	**100**	**25**	**12.26**
Educational level				
Tertiary education	0.0	0.0		
Secondary education	6	6		
Primary education	10	10		
Qur'anic education	77	77		
Adult education	7	7		
No education	0.0	0.0		
Total	**100**	**100**	**25**	**34.70**
Household main source of income				
Sale of fresh product	53	53		
Farming	26	26		
Wage employment	0.0	0.0		
Own business	21	21		
Others	0.0	0.0		
Total	**100**	**100**	**3.33**	**40.57**
Occupation(fishing)				
Primary occupation	59	59		
Secondary education	41	41		
Total	**100**	**100**	**50**	**9.05**
Sold	7	7		
Family consumption	7	7		
Sold and given to family	86	86		
Animal consumption	0.0	0.0		
Total	**100**	**100**	**25**	**30.02**

Source: Field survey (2015)

Fishing gears, methods and crafts used by the fishing communities

Table 2 presents the result for fishing gears, methods and crafts used by the respondents. Passive gears were used by 69% of the respondents such as gill net, hook and line. Seventy one percent (71%) of the respondents engaged in night fishing where they use light and 100% use canoe for fishing.

Processing and preservation methods

Results for the processing and preservation methods used are presented in Figure 2, with forty two per cent (42%) of the respondents selling their fish in fresh form.

Role of women in fishing activities around Sabiyel lake

Most respondents (53%) reported that women engaged in processing and marketing, 28% of the respondents said women engaged in processing and only 3 respondents said women engaged in marketing. Figure 3 present results for the role of women in fishing activities in fishing communities around Sabiyel lake.

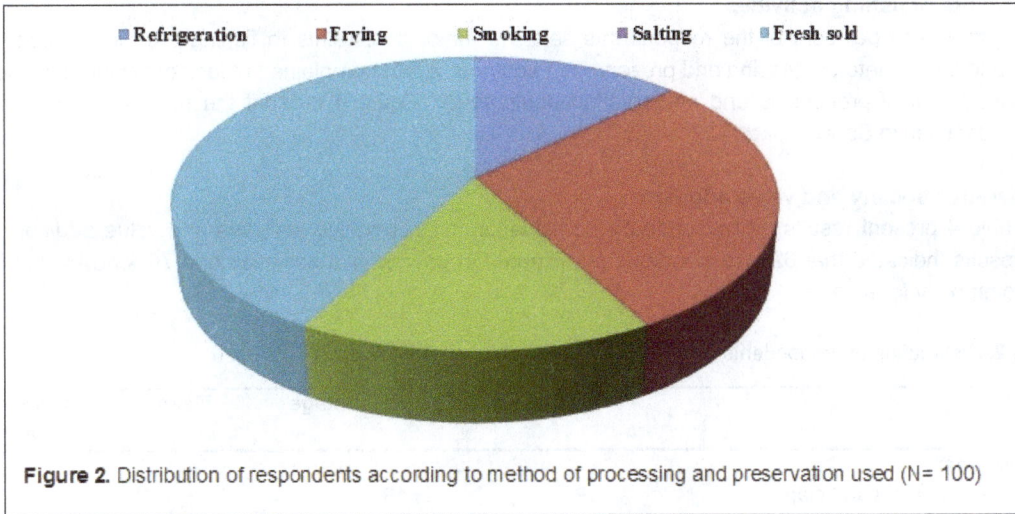

Figure 2. Distribution of respondents according to method of processing and preservation used (N= 100)

Legend: Refrigeration, Frying, Smoking, Salting, Fresh sold

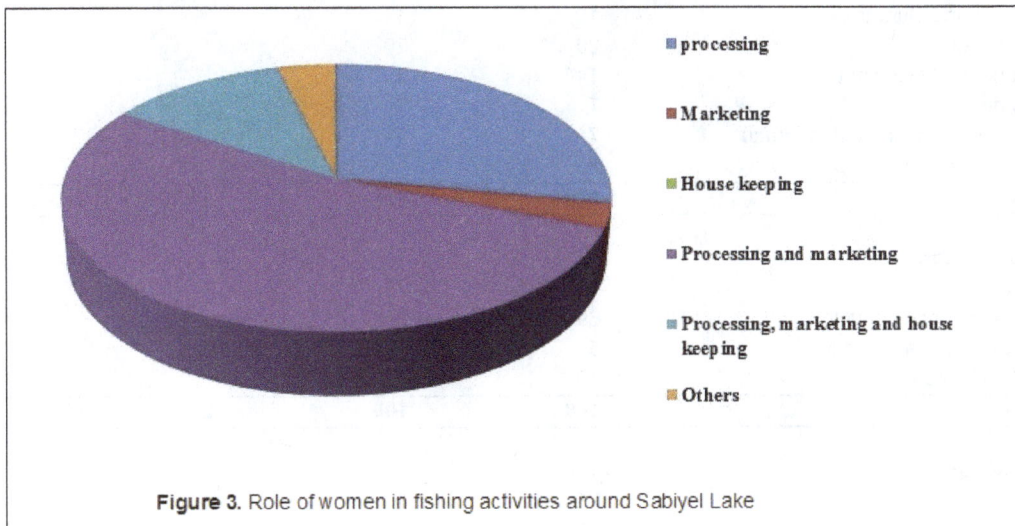

Figure 3. Role of women in fishing activities around Sabiyel Lake

Legend: processing, Marketing, House keeping, Processing and marketing, Processing, marketing and house keeping, Others

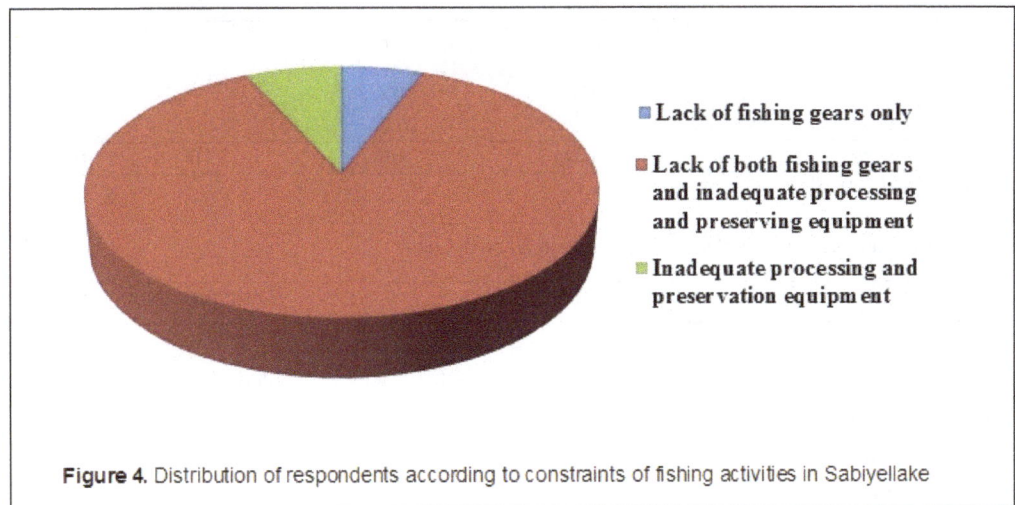

Figure 4. Distribution of respondents according to constraints of fishing activities in Sabiyellake

Legend: Lack of fishing gears only; Lack of both fishing gears and inadequate processing and preserving equipment; Inadequate processing and preservation equipment

Constraints of fishing activities

Eighty seven per cent of the respondents said the major constraints to fishing are lack of both fishing gears and inadequate processing and preservation equipment, 6% complained of lack of fishing gears and 7% mentioned lack of processing and preservation equipment. Figure 4 present the results for constraints of fishing identified in Sabiyel lake.

Cooperative society and value addition

Table 4 present results for respondent's participation in cooperative societies and value addition to fish. The results indicated that 62% respondents participated in cooperative societies and 76% add value to their fish commodity for sale.

Table 2. Distribution of respondents according to fishing gears, methods and crafts used.

Gears	Frequency	Percentage	Mean	Standard Deviation
Active Gears				
Cast net, seine net and clap net	19	19		
Cast net and clap net	13	13		
Cast net and Gura trap	12	12		
Cast net only	20	20		
Cast net and seine net	14	4		
Clap net	7	7		
Seine net, cast net and Induruttu	7	7		
Clap net and Gura trap	8	8		
Total	**100**	**100**	**14.29**	**29.65**
Passive gears				
Gill net	16	16		
Gill net and hook and line	69	69		
Hook and line and Induruttu	5	5		
Hook and line only	10	10		
Total	**100**	**100**	**25**	**880.66**
Use of light				
Yes	71	71		
No	29	29		
Total	**100**	**100**	**50**	**29.69**
Crafts used				
Canoe	100	100		
Others	0.0	0.0		
Total	**100**	**100**	**100**	**0.0**

Source: Field survey (2015)

Table 3. Dominant fish species identified and caught in Sabiyel Lake

Scientific name	Local name	Common name
Heterotis niloticus	Balli	Slap water
Clarias gariepinus	Kullume	Mud fish
Sarotherodon galilieus	Gargazafalga	Tilapia snout
Oreochromis niloticus	Gargazabahausa	Nile tilapia
Hyperopesus bebe	Kuma	Mormyrid

Source: Field survey (2015)

Table 4. Distribution of respondents according to participation in cooperative society and value addition

Cooperative society	Frequency	Percentage	Mean	SD
Yes	62	62		
No	38	38		
Total	**100**	**100**		
Value addition				
Yes	76	76		
No	24	24		
Total	**100**	**100**	**50**	**73.49**

Source: Field survey (2015)

DISCUSSION

The socio-economic characteristics of the fisher folks considered include name of the respondent, village area, household main source of income, marital status, total family size of the fishermen, member of family who go for fishing, amount of fish caught by household which is consumed by the house hold. Because of the overwhelming influence the household head has on the household decision- making process in the traditional Africa setting, most of the socio-economic characteristics discussed were peculiar to the household heads (Ayinla, 2004).

Marital status indicated that 100% of the sampled households were married. This implies that majority of the fisher folks shoulder a lot of responsibilities. Adeleke (2013) also observed that the married were more interested in fishing and attributed that to reliance on fishing to cater for their households. It also shows the availability of family labour in the fishing communities. Age of the respondents vary with 43% of the respondents below the age of 50 years, 32% above 50 years. Eighty per cent of the respondents are below the age of 50 years while only 20% are above the age of 50. Those below 50 years indicated that they were mainly within the economically active age group and physically fit to paddle the canoe. All of the households sampled were male. This implies that the fishing is more popular among men in the study area. This also conforms to FAO (1998) findings that women are rarely involved in fish capture at sea or in lakes because of its inherent dangers, long fishing voyage and their dominant role in household activities.

The fishing gears comprise both active and passive gears. This shows that 19% of the fisher folk were using cast net and clap net, 12% operate with cast net and Gura trap, 16% use gill nets, 16% utilize both gill nets and hook and line, only 5% utilizes hook and line and Induruttu (local fish trap in Nigeria) and 10% uses hook and line, implies that the fisher folks are still using the traditional methods in fishing activities. Among the fisher folks in the study area 71% of the respondents engaged in the fishing activities at night and 29% during the day. This show that the majority of the respondents were engaged in both night and day fishing, which means that they used light during their night fishing and also the main purpose of using the touch light and forehead light is to enable them to see where they set their gears and harvest their fish and also to reset their gears and leave them till in the morning time to come back to harvest again. Only 29% do not engage in night fishing, and also the 100% of the fisher folk household uses paddle canoe for their fishing. This clearly shows that there is a lot of fishing pressure on the lake.

Household heads that had no formal education accounted for 77% of the sampled respondents while the remaining 23% had formal education ranging from adult, primary to secondary education. This implies that very few respondents were educated and only to secondary level. This would have negative consequence on their ability to exploit latent opportunities in fisheries and also to support them in the adoption of improved technologies. Fawole and Fashina (2005) stressed the association of level of education with the use of technology.

In the study area, the 53% of the respondents were involved in only selling of fresh product as a means of livelihood while the other respondents were involved in other income generating activities to boost their income, 26% were engaged in farming activities while 21% engaged in their owned business as their main source of income. Fifty nine per cent(59%) of the respondents had fishing as their primary occupation and

41% had fishing as their secondary occupation. This implies that most households around Sabiyel lake are engaged in fishing with only few engaged in other activities like crop farming. The amount of fish caught by household is partly consumed and partly sold in the market.

The different fish species caught in Sabiyel lake comprises *Hyperopesus bebe, Oreochromis niloticus, Sarotherodon galilieus, Clarias gariepinus* and *Heterotis niloticus.* It was observed that the most dominant fish species in Sabiyel lake are Tilapia and it is more available throughout the year. *Heterotis* and *Clarias* species are also found in Sabiyel lake but they are seasonal, they are more available during rainy season and scarce in dry season. It was also evident that most of the fisher folks sold their fish without processing and preserving. This may be attributed to insufficient processing and preservation methods available in the locality.

Women participated in the fisheries aspect in the study area. According to Adekoya et.al. (2000), artisanal fishing is popular among the men while activities such as processing, storage in the house or housekeeping and marketing are popular among women. It is evident from this research that most women were engaged in both processing and marketing. This agreed with Bene and Merten (2008) who noted that although fish trade is undertaken by both men and women, women have a strong hold on fish trade. According to FAO (2003), fish processing is the exclusive preserve of women, except in some inland fishing communities where men participate actively such as in Lake Chad.

The distribution of major problems confronting the fisher folks in the sampled communities revealed that most of the respondents stated that lack of both fishing gears and crafts are the major problems confronting their fishing activities. The majority (62) of the fisher folk were engaged in co-operative associations and benefited from it. This is supported byILO (2002) that cooperatives are contributing towards gender equality by expanding women's opportunities to participate in local economies and societies in many parts of the world.With regards to value addition, most of the women add value during fish processing to enhance the taste and aroma of fish which increase marketability of fish. To (2015) posited that value addition to fish products enhances better income, improves processing, utilization, keeps in-phase with consumer needs and provides variety of products.

CONCLUSION

The following conclusions were deducted from the study:
- The fisher folks in Sabiyel lake fishing households are still using the crude traditional methods in their fishing activities.
- The majority of respondents within age range of 50 years indicated that they were mainly within the economically active age group and physically fit to paddle the canoe and throwing cast nets.
- Fishing is more popular among men in the study area.
- Most respondents were involved in only selling of fresh product.
- The role of women in artisanal fisheries in Sabiyel lake is mainly in processing and value addition.
- The major constraints to fishing in the study area are the lack of fishing gears, crafts and inadequate processing and preservation equipment.

Recommendation

- Extension services should be intensified to introduce modern methods of fishing with a view to sustainable yield
- The efforts of the economically active age group should be harnessed by educating them on technological advancements in fisheries and aquaculture.
- More fish species should be introduced in to Sabiyel lake to ensure variability and availability while the existing ones should be conserved.
- Fish processing and preservation methods in the study area should be adequately improved.
- Government should address the major constraints to fishing in the study area by supplying and subsidizing fishing gears, crafts and adequate processing and preservation equipment.

COMPETING INTEREST

The authors declare that they have no competing interests.

REFERENCES

1. Abubakar MM and J Auta, 2012. Survey of fisheries resources of Nguru lake. Journal of Biological Science and Bioconservation, Vol. 4:1, pp 8.

2. Adekoya BB, AO Olanuga, OA Adeleye, TO Atansanwo, BG Beckley, AA Idowu and OA Kudoro, 2000. Agro-Economic and Demographic Assessments of the fisher folks in the major Biotopes in Ogun State, Nigeria" Proceeding of the annual National Conference of the Fisheries Society of Nigeria held in Hill Station Hotel, Jos.(19th-24th March, 2000) in press- FISON, Apapa.

3. Adeleke ML, 2013. The socio-economic characteristics of artisanal fisherfolks in the coastal region of Ondo state Nigeria. Journal of Economics and Sustainable Development, 4: 2, p 133-139

4. Ahmad M, FU Shagari and AN Sani, 2014. Fish biodiversity and fishing methods of some water bodies in Katsina state, Nigeria. International Journal of Fisheries and Aquatic Studies, 6: 218-221

5. Araoye PA, 2009. The use of traps and the need for conservation of Synodontis fisheries in Asalake, Ilorin, Nigeria. African Journal of Biodiversity, 8:10, pp 2379-2384

6. Bada AS, 2005. Strategies for bridging the supply-demand gap in fish production in Nigeria, Proceedings of the 19th Annual Conference of the Farm Management Association of Nigeria, pp 329-337

7. Bene C and S Merten, 2008. Women and fish-for-sex: HIV/AIDS and gender in African fisheries. World Development. 36:875-899

8. Eyo AA, 1999. Fish processing and marketing in the arid zone of Nigeria with particular reference to the Lake Chad area and arid zone fisheries. National Institute for Fresh Water Fisheries Research (NIFFR), New Bussa, Nigeria pp 172-175.

9. FAO, 1998. Fisheries Bio-economics Theory, Modeling and management. FAO Fisheries Technical Paper 368. Rome.

10. FAO, 2003. Contribution of fisheries research to the improvement of livelihoods in West African Fisheries communities. Case study:Nigeria. Sustainable fisheries livelihoods programme, SFLD-FAO.

11. (FDF) Federal Department of Fisheries, 2007.Nigeria national aquaculture strategy. Assisted by FAO, Formally Approved by Government, pp 18.

12. Fawole OP and O Fashina, 2005. Factors predisposing farmers to organic fertilizer use in Oyo State, Nigeria. Journal of Rural Economic and Development, 14: 81-90.

13. International Labour Organisation (ILO), 2002. Recommendation 193 concerning the promotion of cooperatives, Geneva: ILO (Available at : http://www.ilo.org/images/

14. Mathew S, 2001. Small-scale fisheries perspectives on an ecosystem-based approach to fisheries management, paper presented at the Reykjavik Conference on Responsible Fisheries in the Marine Ecosystem, Reykjavik, Iceland.

15. Oladimeji YU, Z Abdussalam and MA Damisa, 2013. Socio economic characteristics and return to rural artisanal fisheries household in Asa and Patigi LGAS of kwara state, Nigeria. International Journal of Science and Nature, 4: 445-455.

16. Rahji MAY, TE Aiyelari, OO Ilemobayo and MO Nasir, 2011. An analysis of the agricultural entrepreneurship of broiler farmers in Oyo State, Nigeria. Agrosearch 11:1 & 2, p 83-98.

17. Shimang GN, 2005. Fisheries Development in Nigeria: Problems and Prospects. A presentation by the Federal Director of Fisheries in the Federal Ministry of Agriculture and Rural Development on homestead fish farming training for serving and retiring public servants in the Federal Ministry of Agriculture and Rural Development, FCT, Abuja, p43-45.

18. The FishSite, 2014. Nigeria Begins Import Quota Policy for Fish-News. www.thefishsite.com/fisheries/23029/

19. To JA, 2015. Marketing strategies for value added fishery products. Alsons Aquaculture Corp. Value added fish products-Power Point, 38 pp.

20. Tsadu SM, RO Ojutiku and AV Anyawale, 2006. A survey of fungal contamination of some fish species from Tagwai Daru, Minna, Niger State, Nigeria. Journal of Tropical Biosciences. 6:1-5.

10

POTENTIALITIES OF POND FISH FARMING IN KALIAKAIR UPAZILA UNDER GAZIPUR DISTRICT, BANGLADESH

Md. Mahbubur Rahman[1], Md. Nurunnabi Mondal[1], Jannatun Shahin[2], Jannatul Fatema[3], and Mst. Kaniz Fatema[4*]

[1]Department of Fisheries Management, Bangabandhu Sheikh Mujibur Rahman Agricultural University, Gazipur, Bangladesh; [2]Department of Fisheries, Ministry of Fisheries and Livestock, Bangladesh; [3]Department of Agricultural Economics and [4]Department of Fisheries Management, Bangladesh Agricultural University, Mymensingh-2202, Bangladesh

*Corresponding author: Mst. Kaniz Fatema; E-mail: kanizhossain@gmail.com

---

ARTICLE INFO

Key words

Potentialities
Pond farming
Prospect
Monoculture
Polyculture

ABSTRACT

Potentials and prospects of pond fish farming in improving aquaculture system in Kaliakair upazila under Gazipur district, Bangladesh were investigated. Data were collected from 60 selected fish farmers through questionnaire survey and Focus Group Discussion (FGD) during June to November, 2014. The research revealed that a highest number of pond fish farmers (61.67%) were out of training facilities and a good portion (23.33%) had no education. Pond water was found turbid seasonally (71.66%) and farmers did not exchange water during culture periods (66.67%). As a result water quality deteriorates day by day and depletion of oxygen occurs during pond farming. The average stocking density of fish was higher in the study area and the highest was found in monoculture of Climbing perch (*Anabas testudineus*) (1200 individual/decimal) and lowest in carp polyculture system (195 individual/decimal). Fish production was higher in Pangus monoculture system (17.89 MT/ha/yr) and lower in Climbing perch monoculture system (10.78 MT/ha/yr), but profit was higher in Climbing perch monoculture produced 1318100 Tk./ha/yr and lower in Tilapia monoculture 397886 Tk./ha/yr. Benefit Cost Ratio was higher in Climbing perch (2.32) and lower in Pangus culture (1.34). The problems faced by the fish farmers were broadly categorized as financial, natural, technical and social. Therefore, necessary training facilities with institutional and organizational supports, credit facilities, extension services and awareness development are essential to improve aquaculture system as well as the fishers' livelihoods in Kaliakair upazila of Gazipur district.

INTRODUCTION

The inland fisheries of Bangladesh are one of the most productive resources in the world (Islam and Dewan, 1986). There are about a total of 13 Lac ponds in Bangladesh which covers about 3.05 Lac ha and 2400 km long rivers which covers about 10.32 Lac ha (DoF, 2010). In the year 1998-1999 the total fish production from pond culture system was only 4.99 Lac MT and it was increased 14.47 Lac MT in 2012-2013 year (DoF, 2014). To estimate total number of benefiting people from direct employment in aquaculture and for the improvement of cultural system and future planning, the information regarding present aquaculture practices at the grass root level is absolutely necessary. Survey method is an important necessary way to collect information from bottom level. The area of Kaliakair upazila has huge fisheries resources but its production was lower than other areas like Mymensingh, Narshingdi and even in the average annual fish production in Dhaka division. The river production in the area is decreasing day by day due to decreasing river area and development activities. On the other hand the population of the area was increased to change over time. Although there are huge future prospects of pond fish farming development and to improve livelihood of fish farmer in the area but the information on the said issue is very imperfect. From this point of view, the study was undertaken to determine the potentialities of pond fish farming specially discover the constraints associated with fish farming and the livelihood status of the fish farmers. Thus, the study was initiated with the following objectives:

- Assessment and evaluation of the status and practices of existing pond fish farming in Kaliakair upazila,
- Exploration of the livelihood status and constrain of pond fish farming, and
- Formulation of suggestion and recommendation for improvement and development of pond fish farming in the study area.

MATERIALS AND METHODS

Study area and period

The study was conducted for a period of six months from June, 2014 to November, 2014 in the Kaliakair upazila under Gazipur district of Bangladesh (Map 1). It is very close to the capital city of Dhaka. The site was selected because of the availability of aquatic resources, huge people involvement in fish farming practices, and lack of research activity in this area.

Map 1. Showing the study area

Research design

The design of the study is outlined below:

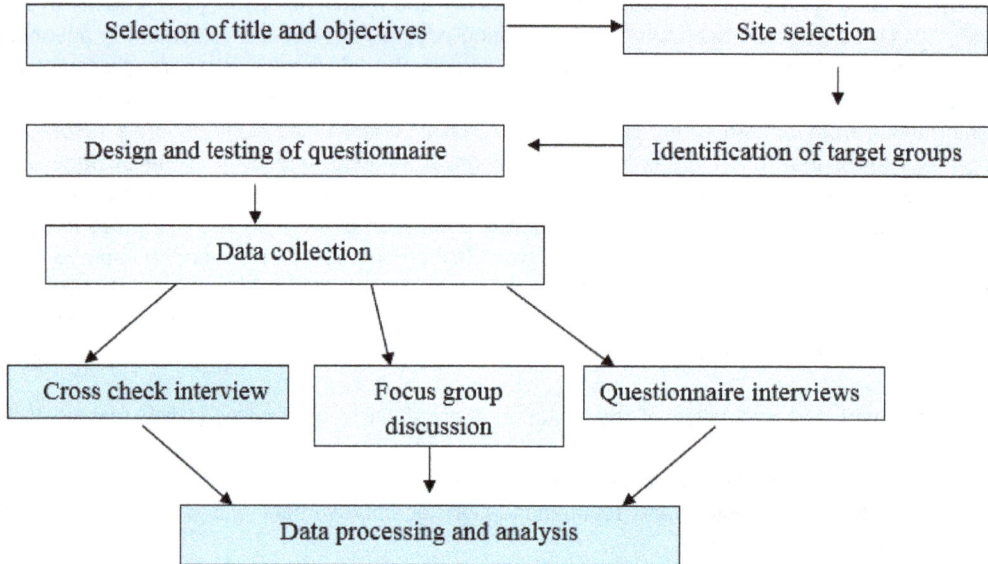

Figure 1. Outline of research activities

Target group
The people who have pond and culture fish were selected as respondents. Data were collected from 60 randomly selected fish farmers covering the selected study area (Figure 1). Most of the farmer culture fish as their primary choice to support their family and to improve their social condition.

Design and test of questionnaire
A set of questionnaire interview schedule was designed. The draft questionnaire was tested with 10 fish farmers in the study area. In the pilot survey, much attention was given to any new information which was not designed but was important and informative towards the objective. The questionnaire was changed, modified and rearranged according to the experience gathered in pre-testing of questionnaire. The final interview schedule was then developed in logical sequence so that fish farmers could answer systematically. Data were collected by direct interview using questionnaire and cross check interview. Fish farmers were interviewed at their house or farm sites and the information was recorded by the researcher himself.

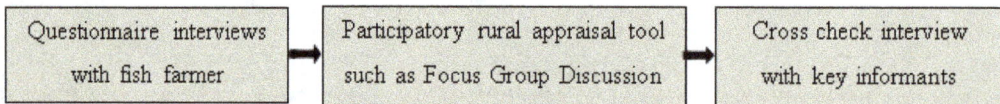

The data were collected through questionnaire interviews and FGD, and the cross check interviews were conducted with key informants such as Upazila Fisheries Officer (UFO) and relevant NGO workers.

Data processing and analysis
All the collected data were summarized and scrutinized, and analyzed and recorded carefully. Finally the relevant tables were prepared in accordance with the objectives of the study. Data were presented mostly in the tabular form because it was simple in calculation and easy to understand.

RESULTS AND DISCUSSION

Demographic profile

According to the census of Bangladesh population in 1981, 1991, 2001 and 2011 the total populations of the upazila were 165766, 232915, 267003 and 483308, respectively. This showed that the population of the area was increased in a dramatic rate. The total population in the census 1981 was 165766 that reached 483308 in 2011 census. The rate of increase of male population was higher than female population in every census.

Fisheries resources

Ponds, rivers, beels and ditches were found as water resources in the study area (Table 1). Most of the areas were related with fish culture and agriculture.

Table 1. Fisheries resources in Kaliakair upazila

Serial	Resources	Number/area
1	Total land area	314.14 sq.km
2	River	3
3	River area	2824 ha
4	Flood plain area	2833 ha
5	Pond	2852
6	Pond area	1017 ha
7	Fish farmer	2935
8	Fishermen	1514
9	Fish hatchery	1
10	Beel nursery	2311
11	Feed mill	4
12	Fish Suntuary	22

Source: Citizen Charter, Upazila Fisheries Office, Kaliakair, Gazipur

River and its present status in Kaliakairupazila

Three rivers name Turag, Bongsi and Gualia are flow through the Kaliakair upazila. The total river area of the upazila is gradually decreasing due to political and local tenant, continuous industrial development, construction of roads and highway, unplanned growth of living areas, building of market in the bank of river. In 2008, the total river area was 3234 ha but it decreased in 2015 and reached in 2824 ha. Fish production in the river is also decreasing. Therefore, to meet the demand and to supply necessary fish protein for the increasing people, it is urgently needed to develop the aquaculture system of the area.

Socio-economic condition and livelihood status of pond fish farmer

From the study it was found that about 41.67% monoculture farmers were middle age (36-45 years), 16.67% young and 5% old. The result of the study revealed that young and middle age farmers were more involved in monoculture farming, on the other hand middle and old age farmers involved in polyculture farming. Ali et al., (2008) studied on socio-economic condition of small farmer and reported that most of the farmers were belonged to the age category of 31 to 40 years. About100% monoculture farmer and 90.09% polyculture farmer was male that means that no females involved in monoculture fish farming systems. But if women can be involved in fish farming activities, they can utilize their leisure period simultaneously support farming activities and earn money. Zaman et al., (2006) studied on the assessment of livelihood status of fish farmers in some selected areas of Mohanpur upazila under Rajshahi district and found that 85% male and 15% female farmers were involved in pond fish farming activities. Religious status of fish farmer was 95% Muslim and 5% Hindu. Family size of monoculture farmer (36.66%) was medium size (6-9 members) and 10% family of polyculture farmer was small size. In surveyed area about 56.67% farmer received fish farming as their main occupation and also found 10% service, 20% business, 13.3% agriculture as their main occupation.

The education level of the fish farming community in the study area was lower (Figure 2). In case of polyculture pond fish farming the majority of the respondent (18.33%) was in illiterate. The mean literacy was found 8.56±2.24 in monoculture and 6.91±1.3 in polyculture farming system. Among the respondent a major portion (23.33%) had no education that they were illiterate and the highest portion (25%) had only secondary level of education. The literacy of the fish farmers is an important factor, which determines their communication behavior, access to the printed and mass media as well as acquainted with the local and world market.

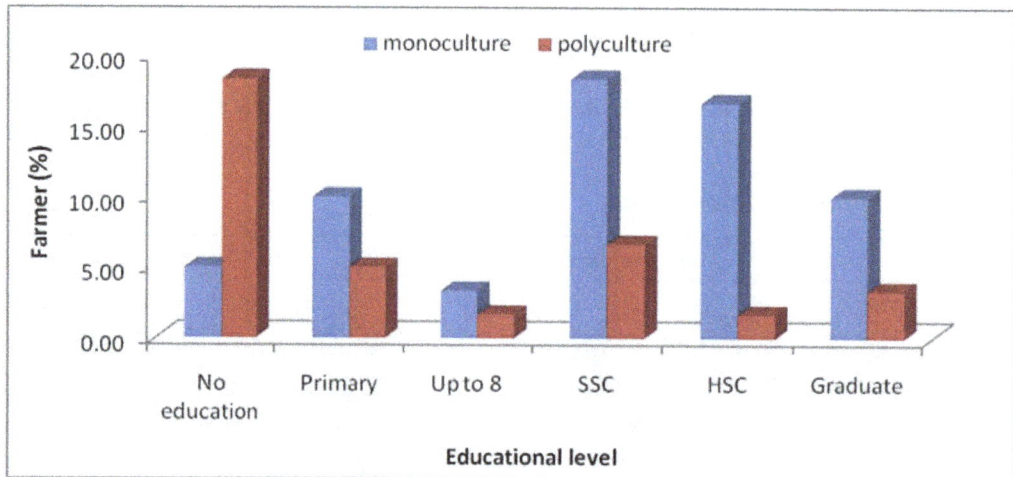

Figure 2. Educational status of farmer in the study area

Rahman (2012) studied an experiment on socio-economic condition of commercial climbing perch fish farmer in Muktagacha uapzila, Mymensingh district and found that about 28% of the fish farmers had literacy up to primary level, 4% were illiterate, and 46% up to secondary level and 14% could sign only. The study revealed that the sanitary conditions of the fish farmer were not satisfactory in the study area. Sanitary facilities of polyculture farmer was very poor about 43.33% had non-constructed facility. Kabir et al., (2012) found in their study that 60% of the farmers had semi-constructed, 30% non-constructed and 10% had no sanitary facilities in their study area. A significant proportion of fish farmers (55%) had taken their health service from village quack doctor (who actually possess no knowledge on medical science) and it was found that 43.33% fish farmers were living in tin shed house (non-constructed). It was observed that 68.33% fish farmers had no own land, they culture fish in lease pond and the famers spent their maximum profit for paying the lease pond. Bank loan was higher in monoculture farmers about 33.33% received loan for fish culture and a total of 36.67% respondents invest their own money for fish culture. The involvement of NGO in monoculture farming system was very minor only 5%. Mean annual income of monoculture farmers were 556972 BDT and polyculture farmers 27022 BDT. The polyculture farmers usually do not consider farming as a business. As their farms are not commercial in nature, they do not seriously think about profit and loss.

Status of pond fish farming system

Physical structure and condition

Pond ownership

Surveyed ponds were classified into four categories on the basis of ownership. In monoculture farming system, it was observed that 18.33% ponds under single ownership, 41.67% under multiple ownerships with 2-3 owners and the rest 3.33% ponds were public property. Ownership pattern of polyculture ponds were observed that 23.33% of total surveyed pond under single ownership, 11.67% under multiple ownerships ranging from 2-3 owners and the rest 1.67% ponds were public property or property of some organization. Ali et al., (1982) studied the ponds ownership of six upazila's of Mymensingh district and found that 84% under joint ownership and only 16% under single ownership. From this discussion it was observed that researcher studied on pond ownership found that majority of pond under joint owner. This is also similar to the findings of present study.

Farm size

Farm size is an important variable for the production of fish (Islam and Dewan, 1986). In the study the respondents were classified into three categories according to their farm size and their farm size ranging from 0.3 to 1.7 ha. The mean (±SD) farm size of the monoculture and polyculture farmers were 0.45±0.05 ha and 0.64±0.01 ha, respectively.

In monoculture farm, about 20% of the respondents had small size farm and 36.67% had medium size and only 6.67% had large size farm. It was found that the highest percentage (36.67%) of respondent had medium size farm and involved in single species culture activities and lowest had (5%) larger size farm and involved in multi-species culture system. The farmers had very small size farm, they didn't earn enough profit from fish culture. On the other hand a major portion of large farm size farmers didn't culture fish in scientific methods, so, they also not earn enough money from fish culture. Rahman (2003) found in his study that the average farm size was 0.12 ha with a range from 2.50 to 15.0 ha in Gazipur district. Saha (2003) observed that the range of farm size were within 0.05 to 0.15 ha in his study.

Water quality condition

Depth and source of pond water

The water depth was found 3-5 ft, 5-8 ft and above 8ft in the category of small, medium and large, respectively. Among the study the water depth of 36.67% monoculture ponds and 23.33% polyculture ponds were 5-8 feet (1.52-2.44 m) during rainy season. According to Jone and Paul (2012) the depth of ponds is generally in the range of 0.8 to 1.8 m (2.62-5.90 ft) these depths allow adequate light penetration for primary productivity. In the study water depth was higher than recommended depth for pond fish culture. It was found that about 46.67% monoculture farmer supply water in their pond from ground water source by using deep tube well and 33.67% polyculture farmer depend on rain and river water as a source of pond water. In the study it was found that a large number of farmers (55%) depend on rain and ground water for fish culture in their ponds. During the time when rainfall is low, they totally depend on ground water. As a result increased their production cost and simultaneously decreased their profit.

Water exchange

In the study, about 46.67% monoculture farmers and 20% polyculture farmers did not exchange water during culture periods. Rahman (2007) studied on pond fish farming and livelihoods of rural fish farming in some selected areas of Kurigram district and observed that farmers have no facilities to exchange water. From the study it was found that 66.67% farmers did not exchange water during culture periods from their ponds. But during culture of fish, large amounts of metabolites were continuously excreted and deposited into the pond bottom and moreover, where excess, unconsumed feeds also added to the bottom load and serve to pollute the water. To prevent the deterioration of the pond environment, pond water is continuously freshened by the entry of new water from the water source, while old water is drained through the outlet/drainage gate.

Turbidity and water colour

Turbidity is the degree of opaqueness produced in water by suspended particulate matter. Intensity of turbidity varies with soil type, season, amount of surface runoff, amount of organic decomposition and others. In the study three types of water color observed in farmers' ponds. Water colour of monoculture farm was 43.33% muddy, 13.33% brown, 6.67% light green and 28.33% muddy, 6.67% brown and 1.67% light green in polyculture farm. Dosdat et al., (2001) studied on the environment impact of aquaculture and found that water colour affects different element in pond. In the study it was found that about 71.66% ponds water was turbid seasonally with clay or soil particles.

Measurement of water quality

Production of sufficient fish food organisms highly depend on the water quality. In the study it observed that all farmers were not able to measure the important physical and chemical parameter (Table 2) of pond water due to lack of instrument, high cost of equipment, lack of technological knowledge and high cost associated with water quality measurement. The farmers were measured some of the physico-chemical parameters of their ponds with the help of upazila fisheries office and LEAF (Local Extension Agent for Fisheries). Hasan and Ahmed (2001) studied on issues in carp hatcheries and nurseries in Bangladesh, with special reference to health management and aquaculture development and found that some of the rural small scale farmers observed water quality parameter of their culture pond.

Table 2. Water quality parameter measured by the farmers in the study area

Parameter	Monoculture		Polyculture	
	Respondent	Percentage (%)	Respondent	Percentage (%)
Temperature	12	20	6	10
pH	3	5	-	-
Dissolved oxygen	3	5	2	3.33
Transparency	30	50	14	23.33
Alkalinity	2	3.33	-	-
Ammonia	-	-	-	-
Phosphorus	-	-	-	-
Nitrate	-	-	-	-
Chlorophyll-*a*	-	-	-	-

Sources of fish seed

In Kaliakairupazila had only one private hatchery and no government hatchery. The hatchery produced only carp seed and not full fills the farmers demand. It was observed that only 15.33% farmers' got fish seed from private hatcheries of Gazipur district. Therefore, to full fill the demand, farmers collect fish seed from the hatcheries of Mymensingh, Bogra and Rajshahi district (Table 3).

Table 3. Sources of fish seed in the study area

Sources	Monoculture		Polyculture	
	Respondent	Percentage (%)	Monoculture	Percentage (%)
Private hatchery (Mymensingh)	21	12	43.33	20
Private hatchery (Rajshahi)	6	1	10	1.67
Private hatchery (Bogra)	8	4	13.33	6.67
Private hatchery near farm	3	7	5	8.33
Natural sources	0	0	0	0

Stocking density

Stocking activities depends on supply and availability of fish seed. Most of the farmers stocked fish fry/fingerling in the month of June-July when the pond had accumulated about 5-8 feet of rain water. Farms with a perennial water source were stocked as early as the month of April-May. Generally, farmers were released of fish fingerlings to ponds in around June and cultured as long as sufficient water retained in the pond. Stocking density of monoculture (Tilapia), monoculture (Pangus), monoculture (Climbing perch) and polyculture (Indian major carp) were 48782 individuals/ha, 58508 individuals/ha, 288166 individuals/ha and 43225 individuals/ha, respectively (Table 4).

Table 4. Stocking density of fish fingerling in the study area

Culture strategy	Stocking density/decimal	
	Mean	SD
Monoculture (Pangus)	236	56.35
Monoculture (Climbing perch)	1165	242.21
Monoculture (Tilapia)	197	52.82
Polyculture (Indian major carp)	175	23.24

According to Islam (2012) the average stocking density of fish better for pond aquaculture was 17,370 fry/ha/year. Parvin (2011) reported that average stocking density in case of Pangus 32,000-45000 fingerlings/ha and in case of Climbing perch 80,000-90,000 fingerlings/ha in three upazila of Mymensingh district. In the study area the lower (mean ± SD) stocking density was 175 ± 23.24 fingerlings/decimal in polyculture system, while the higher (mean ± SD) stocking density was 1165 ± 242.21 fingerlings/decimal of Climbing perch in monoculture system. The average stocking density of fish was higher in the study area and in case of monoculture Climbing perch it was very high. It means that "More stocking gives more production" was the main idea of farmer in the study area. Sometimes they denied the suggestions of UFO, specialist or extension personnel.

Feeds and feeding strategy

In the study areas farmers mainly used three types of feed such as homemade feed (locally called loose feed), commercial feed and homemade feed both (loose + pellet). Farmers prepared feed by their own feed machine. Farmers used rice bran, wheat bran, mustard oil cake, fish meal, bone meal and vitamin-minerals premixed as major ingredients for the preparation of homemade feed. Among the commercial feed, farmers frequently bought feed from Quality, Saudi-Bangla, ACI feed, Aftab feed and Mega feed company. Price and quality differs from one company to another and within the same company. In case of monoculture farming practice, about 43.33% farmers used both types of homemade feed (loose + pellet), 6.67% farmer used only homemade feed (loose) and while 13.33% farmers used commercial feed. Farmers applied feed at an average or 4.5% body weight with a range of 3-5% body weight. Farmers reported the FCR (Food Conversion Ratio) value ranged from 1.5 to 2.0 with different feeds. Rasel (2011) found in his studied that about 85% farmers used commercial feed and 15% used homemade feed in Tilapia farming in Mymensingh district. In case of polyculture system, the study reveals that 28.33% farmers used commercial pellet, 1.67% used both type of homemade feed (loose and pellet), 6.67% farmer used only homemade feed (loose). In the study, maximum farmers used rice bran because it is available and low price. Farmers generally used different hormones, antibiotics and growth promoter that could be sometimes harmful.

Use of chemicals, drugs, antibiotics and toxic substance

In the study area it was found that all (100%) respondents used lime, 90% of monoculture farmer used Zeolite, Panvit aqua, Zeofresh, Gasonex and Biomax in their pond by the instruction of company agents. About 10% farmers used $KMnO_4$, 75% used antibiotics, 2% used Dipterex, 8% used copper Sulphate, 5% used Malachite green, 5% used Methylene blue and 3% farmers used Calcium hypochlorite when disease problems appear. It also found that 82% of monoculture farm used antibiotics. Aoki (1992) reported that the use and sometimes abuse of antibiotics in more intensive farming led to multiple drug resistance among pathogens. Pillay (1992) stated that there is a possibility of generating drug-resistant strains of pathogens by the use of antibiotics for treating diseases into the environment.

Production of fish

In the year 2013, annual yield of monoculture Tilapia, Pangus and Climbing perch were 15.8 MT/ha/yr, 16.95 MT/ha/yr and 10.12 MT/ha/yr, respectively and the polyculture of Indian major carp was 11.86 MT/ha/yr in Kaliakair upazila (Table 5). Parvin (2011) found the average yield of Pangus was 25,811 kg/ha and Climbing perch 53,350 kg/ha in three upazila of Mymensingh district.

Table 5. Fish production, cost, revenue, profit, and BCR per hectare in the year 2014 in the study area

Species	Production (MT)	Cost (Tk.)	Revenue (Tk.)	Profit (Tk.)	BCR
Tilapia	16.27	1081656.88	1479543.33	397886.45	1.37
Pangus	17.89	1257208.98	1690066.23	432857.25	1.33
Climbing perch	10.78	996741.0565	2314841.92	1318100.86	2.32
Indian major carp	12.16	1650856.957	2487573.74	836716.78	1.51

The highest species wise fish production was Pangus (17.89) and lowest was in Climbing perch (10.78). The above discussion indicated that the fish production was increased in the study area but it is fur from the national target and it is possible to further increase. In the year 2012-2013 the annual fish production of pond in Kaliakair upazila was 2.99 MT/ha but the national annual fish production in pond was 3.89 MT/ha (FRSS, 2014). According to FRSS, 2014 the annual fish production of pond was 5.48 MT/ha in intensive pond fish farming and it was increased up to 22.70 in highly intensive pond fish farming. In the year 2013-2014 the annual pond fish production in Kaliakair upazila was 3.80, but the average national annual pond fish production was 4.1 (Azad, 2015). According to Azad, 2015 if all the pond of the country will be taken under sustainable aquaculture through the extension of appropriate technology then it will be possible to produce 5.0 MT/ha of fish in the pond within 2020-2021.

Gross and net return from pond fish farming

In the study, it was found that the average annual return from the production of Tilapia, Pangus, Climbing perch and Indian major carp were 1479543.34 Tk./ha/yr, 1690066.23 Tk./ha/yr, 2314841.92 Tk./ha/yr, 2487573.74 Tk./ha/yr, respectively. Rahman (1995) observed that the average gross and net returns of carp were 72,910 Tk./ha/yr and 15,833 Tk./ha/yr, respectively in Tarakanda upazila of Mymensingh district. In the study it found that the highest return was in polyculture of Indian major carp and lowest in Tilapia farming.

Benefit Cost Ratio (BCR)

In the present study it was found that the average BCR in monoculture pond farming of Tilapia, Pangus and Climbing perch were 1.37, 1.33 and 2.32, respectively and in polyculture pond farming of Indian major carp was 1.51. Awal et al., (2001) was estimated the overall economic return (net return) and BCR of Pangus culture as 23964 Tk./ha and 2.73, respectively in Jamalpur and Sherpur District. Sohag (1996) found in his studied that the BCR of Tilapia was 2.02 in Nandail Thana Mymensingh district. From the study it observed that the highest BCR (2.32) was in Climbing perch and lowest (1.33) in Pangus farming. The study also revealed that the benefit was higher in monoculture Climbing perch farming compare to polyculture of Indian major carp.

Harvesting and marketing of fish

The farmer intermittently harvested fish for family consumption or at 1-2 times for marketing. The peak period of harvesting was September to November month. Most of the farmers (64%) practiced total harvest and others (36%) practiced partial harvest in the month of August to November for selling. Farmers harvest their fish by using cast net and seine net or by total drying of pond. It was found that about 70% of fish sold to the wholesalers or local agents for transportation to the Dhaka city and the rest (30%) sold for local retail market. The harvested fish reached from culture pond to consumer by three different ways (Figure 3). Farmers reported that they were facing some problems during marketing due to narrow muddy road, lack of transport facilities and poor marketing system.

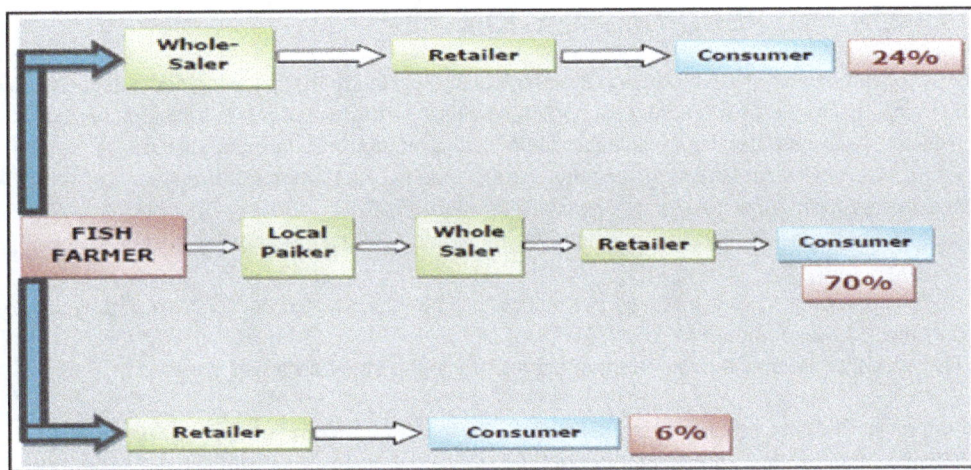

Figure 3. Marketing channel of fish in Kaliakair upazila

Constraints of pond fish farming in the study area

In the study area had only one private hatchery and no government hatchery, and the existing hatchery produced only carp seed. On the other hand the hatchery not full fills the demand of fish seed. So, to meet up the demand the farmers collect fish seed from a long distance according to availability, these increase the cost of production. The pond fish farmers in the study area were also facing various problems during culture of fish. These problems broadly categorized as financial, natural, technical and social. The farmers confronted the problems during pond fish farming were ranked and index in the following Table (Table 6).

Table 6. Rank order of problems in monoculture farming in the study area

Problem	Score of extent of problem confrontation					
	H	M	L	N	PCI	Rank
Inadequate supply of fish seed/fry/fingerling	84	21	9	0	109	1
Lack of finance	78	20	6	0	104	2
Low growth rate of fish	75	21	6	0	102	3
High prices of fish feed	75	18	6	0	99	4
Low quality of feed	69	`20	9	0	98	5
Water sources to fill up pond	63	22	10	0	95	6
Mortality of fish	60	20	10	0	90	7
Water quality deteriorated	60	20	9	0	89	8
Industrial pollution	54	22	11	0	85	9
Training facilities	51	18	10	0	78	10
Availability of preservation (ice) materials	48	14	12	0	73	11
Multiple ownership	42	12	11	0	68	12
Poaching of fish	36	12	10	0	61	13
Availability of manpower	33	14	7	0	54	14
Poor marketing facilities	30	16	9	0	50	15
Political problems	24	16	8	0	48	16
Fertilizer and manure application	18	14	12	0	44	17

H = High, M = Medium, L = Low, N = Not at all, PCI = Problems Confronting Index
Here, PCI = (H*3+M*2+L*1+N*0)

CONCLUSION AND RECOMMENDATION

Based on the major findings of the study and their logical interpretation the following conclusions were drawn:
- Farming practices of monoculture and polyculture farmers were not satisfactory due to lack of sufficient fish seed, training facilities and knowledge on intensive farming system.
- Pond size, fry size, stocking density, water quality, embankment condition of pond was not satisfactory for monoculture and polyculture farming.
- Indiscriminate use of feed, chemicals, antibiotics and fertilizer decreasing sustainability of pond fish farming.
- No female member were involved in monoculture farming system and in case of polyculture system only few female members were involved.
- The education level of the fish farming community in the study area was lower.

Based on the major findings, problems and conclusion the following recommendations were made:
- Women and young age people could be more involved in monoculture system to increase aquaculture production.
- Regular checking of water quality parameter should be made and a control measure should be taken against the indiscriminate use chemical and drug.
- Natural and artificial water reservoirs should be constructed for supplying water during dry season.
- Educational institution should be set up to improve educational status.
- Government, private sector and NGOs should come forward to establish fish hatchery and fish processing plant.
- Government and other institution should provide sufficient fund and facilities.

REFERENCE

1. Ali H, MA Akber and MH Rahman, 1982. Utilization of fish production in Mymensingh District. Bangladesh Journal of Agricultural Economics, 5: 103-114.
2. Ali MH, ANGM Hasan and MA Bashar, 2008. Assessment of the livelihood status of the fish farmers in some selected areas of Bagmara upazila under Rajshahi District. Bangladesh Journal of Agricultural University. 6: 367-374.
3. Aoki T, 1992. Chemotherapy and resistance in fish farms in Japan. In: Diseases in Asian Aquaculture, Asian Fisheries Society, Manila, 1: 519-529.
4. Awal MA, MA Ali and MGF Mia, 2001. Fisheries resources and utilization pattern in some selected areas of Jamalpur and Sherpur district. Bangladesh Journal of Training and Development, 14: 183-190.
5. Azad SA, 2015. Fisheries sector in socio-economic development of Bangladesh. National Fish Week 2015 compendium (In Bengali), Department of Fisheries, Ministry of Fisheries and Livestock, Bangladesh. Pp. 13-23.
6. DoF, 2010. National Fish Week 2010 compendium (In Bengali), Department of Fisheries, Ministry of Fisheries and Livestock, Bangladesh, Pp.102-103.
7. DoF. 2014. National Fish Week 2014 compendium (In Bengali), Department of Fisheries, Ministry of Fisheries and Livestock, Government of the Peoples' Republic of Bangladesh. 13 P.
8. Dosdat A, AU Vilalba and B Basurco, 2001. Environment impact of aquaculture in Mediterranean: nutritional and feeding aspects. In: Proceedings of the seminar of the CIHEAM Network on technology of aquaculture in the Mediterrean (TECAM), jointly organized by CIHEAM and FAO, Zaragoza, Spain, Vilalba, A. U. (ed.) Pp. 23-36.
9. FRSS, 2014. Fisheries Statistical Yearbook of Bangladesh.Fisheries Resources Survey System (FRSS), Department of Fisheries, Bangladesh, 30: 52 P.
10. Hasan MR and GU Ahmed, 2001. Issues in carp hatcheries and nurseries in Bangladesh, with special reference to health management. In: Primary Aquatic Animal Health Care in Rural Small-scale Aquacultural Development, Arthur JR, Phillips MJ, Subasinghe RP, Reantaso MB and MacRae (eds.), FAO, Fisheries Technological Paper. 406: 147-164.

11. Islam MS and S Dewan, 1986. Resources use and economic return in pond fish culture. Bangladesh Journal of Agricultural Economics, 9: 141-150.

12. John SL and PC Southgate, 2012. Aquaculture: farming aquatic animals and plants. Second edition. Wiley-Balckwell publishing Ltd. Pp. 648.

13. Parvin S, 2011. Present status of commercial aquaculture in three upazila of Mymensingh district, MS Thesis, Department of Aquaculture, Bangladesh Agricultural University, Mymensingh. Pp. 21-22.

14. Pillay TVR, 1992. Aquaculture and the environment. Fishing News Book, Blackwell Scientific Publications Ltd., Osney mean. Oxford OX2 OEL., England. 189 P.

15. Rahman MM, 2007. Studies on pond fish farming and livelihoods of rural fish farming in some selected areas of Kurigram district, MS Thesis. Department of Aquaculture, Bangladesh Agricultural University, Mymensingh. Pp. 67-68.

16. Rahman MM, 2012. Socio-economic condition of commercial Koi fish farmer in Muktagacchaupazila under Mymensingh district, MS Thesis, Department of Aquaculture, Bangladesh Agricultural University, Mymensingh, Pp. 25-26.

17. Rahman MM, 1995. An economic study of pond fish culture in some selected area of Mymensingh District. An M.S. thesis submitted to the Department of Agricultural Economics, Bangladesh Agricultural University, Mymensingh, 94 P.

18. Rasel MMH, 2011. Socio-economic study of Tilapia farmers in Mymensingh region.MS Thesis, Department of Aquaculture, Bangladesh Agricultural University, Mymensingh. Pp. 43-44.

19. Saha MK, 2003. A Study on Fish Production Technology in North-West Bangladesh.MS Thesis, Department of Aquaculture, Bangladesh Agricultural University, Mymensingh. 45 P.

20. Shohag, MS, 1996. An Economic Study on the Supervised Credit Pond Fish Culture in Nandali Thana of Mymensingh District. MS Thesis, Department of Agricultural Economics, Bangladesh Agricultural University, Mymensingh. 24 P.

21. Zaman T, MAS Jewel and AS Bhuiyan, 2006. Present status of pond fishery resources and livelihood of the fish farmers of Mohanpurupazila in Rajshahi District. Bangladesh Journal of Fisheries Research, 25: 31-35.

11

LENGTH-WEIGHT RELATIONSHIP AND CONDITION FACTOR OF BOMBAY DUCK *Harpadon nehereus* FROM LANDINGS OF FISHERY GHAT, CHITTAGONG

Tasnuba Hasin

Department of Fisheries Resources Management, Faculty of Fisheries, Chittagong Veterinary and Animal Sciences University, Chittagong-4225, Bangladesh

*Corresponding author: Tasnuba Hasin; E-mail: shaily.1119@gmail.com

ARTICLE INFO	ABSTRACT

Key words

Harpadon nehereus, Length-weight relationship, condition factor, Allometric growth

The current study describes the length-weight relationships and condition factor of Bombay duck, *Harpadon nehereus* from landings of Fishery Ghat, Chittagong. A total of 300 individuals caught by various mesh size of gill nets by commercial fishers were investigated from April 2016 to June 2016. On the basis of hypothetical cube law $W=aL^b$, the variance of 'b' for both total length and standard length were estimated statistically (0.0090 and 0.0088 respectively). Using the value of t=1.96 at the 95% confidence limits for 'b' were estimated as 2.218 to 2.590 for total length-weight relationship and for standard length-weight relationship, 1.985 to 2.353 which indicated allometric growth pattern for both cases. The monthly relative condition factor during three months of observation (0.978, 1.033, and 1.097) showed no notable difference due to spawning strain, spent condition and lower feeding rate right after peak spawning season.

INTRODUCTION

The Bombay duck *Harpadon nehereus* lives in the tropical waters of the Indo-Pacific- and traditionally caught in the waters of the Bay of Bengal. According to the Department of Fisheries (2005-2006) Bombay duck (Synodontidae) contributed 8.20 percent (37 331 t) to the total marine fish catch.Bombay duck is a soft fish and is highly perishable because of its body composition. A large part of the catch particularly during the peak fishing season is sundried on raised bamboo platforms by hanging them on ropes and it is an important exported fishery product.

In Bangladesh, the species is found mainly in the Bay of Bengal and estuaries of Bangladesh and also ascends tidal rivers (Rahman, 1989 and 2005). In addition, they are inhabitants of saline and semi saline waters of the Sundarbans (Gopal and Chauhan, 2006, Hoq et al., 2006).It is abundant in the Northwest coast of India (Rupshankar,2010).This fish is of commercial significance (Rahman, 1989 and 2005) and used as food fish in Bangladesh both fresh and sun dried. According to Rahman (1989 and 2005) it is highly esteemed as food particularly in Chittagong (Bangladesh) where these are found in abundance. He also mentioned that dried Bombay duck has commercial importance in the South and Southeast Asia. In India, the fish contributes a substantial fishery in the Hooghly estuary (Talwar and Jhingran, 1991). However, Kamal et al. (2001) treated this species as under-utilized fish in Bangladesh.

Due to initiation of mechanization of fishing crafts, the seaward fishing activity of the Chittagong coast has considerably increased in recent years. The abundance of Bombay duck along the coast is gradually increasing. Bombay duck fishery is under constant pressure due to various stresses such as habitat destruction because of pollution, over-exploitation, indiscriminate killing of juveniles etc, which in turn help the decline of this fish population. Therefore, there is an immense need to manage the fisheries more vigilantly to ensure sustainable fish production in future. In this regard fish stock assessment plays an important role in the rational management and conservation of this resource. The population Dynamics and stock assessment facilitate to know about fish growth, mortality rate, spawning time, catch rate, maximum sustainable yield, maximum economic yield etc (Larkin, 1977). The mesh size regulation, implementation of ban at breeding season and selective fishing are the results of the stock assessment of the fishes. The present study was intended to investigate the length-weight relationship and condition factor of *H. nenereus* which is an essential part of growth study and overall health status. These data are of important scientific findings to determine how fast the fish grow and the recovery time of its population after exploitation.

MATERIALS AND METHODS

Collection of data

The present study was conducted in Fishery Ghat fish market, Firingi Bazar, Chittagong, Bangladesh during April 2016 to June 2016 of which sampling was carried out at monthly interval. The samples were collected from different commercial gill-netters. They were transported to laboratory for length and weight data measurement. The total length (from the tip of the snout to the end of the caudal fin) and standard length (from the tip of the snout to the mid- of point caudal peduncle) of 300 fish were measured using a meter scale (1±mm) and weighed to the nearby 'g' using precision balance (0.001g). Total and standard length varied in the size range of 11.8 cm to 25.4 cm and 9.9 cm to 21.4 cm, correspondingly, and the weight ranged from 4.36 g to 52.71 g.

Analysis of Data

Length-weight relationship

The length-weight relationship of *Harpadon nehereus* was estimated using the power curve equation, $W = aL^b$(King, 2007) where,
W= Body weight (g) of the fish,
L= Total length (cm) (TL) or standard length (cm) (SL) of fish and 'a' and 'b' are the constants.

The value of 'b' is used to estimate the individual growth, in which hypothetically following a premise that b = 3 stands for isometric growth (the length growth in line with the increasing weight) whereas b > 3 indicates that the weight gain is faster than length (positive allometry) and oppositely b < 3 points to negative allometry.

The above equation can be transformed into a linear form using natural logarithm as lnW = ln a + b ln L where Ln a and b equate to the intercept and slope. A power curve-of-best-fit (Figure 1a and 1b) for both total length and standard length is drawn through the points representing predicted weight for a range of arbitrarily chosen values for length wherein the value of 'a' and 'b' and the chosen value of L in the power equation were substituted (Table 1). The non-linear and linear equations were fitted separately for total length (TL) and standard length (SL). The correlation coefficient (r) and variance of 'b' (S_b^2) were calculated following standard statistical procedures. Using the value of 't' from statistical table (King, 2007)with n-2 degrees of freedom the 95% confidence limits are estimated as b ± t× S_b.

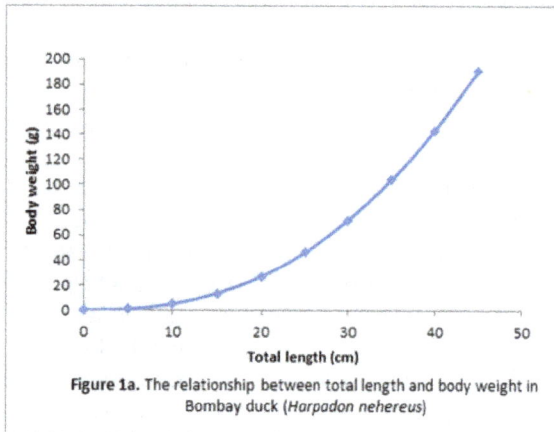

Figure 1a. The relationship between total length and body weight in Bombay duck (*Harpadon nehereus*)

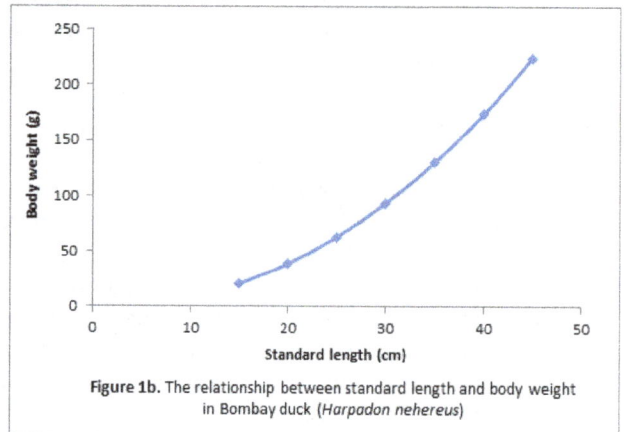

Figure 1b. The relationship between standard length and body weight in Bombay duck (*Harpadon nehereus*)

Relative condition factor (CF$_w$)

The relative condition factor (Kn) of *H. nehereus* was calculated using the formula CF$_w$= $\overline{W}$/W$_{pred}$ (King, 2007) based on total weight relationship,

where CF$_w$= relative condition factor;

$\overline{W}$ = mean weight and

W$_{pred}$ = predicted weight.

Monthly values of mean weight are compared with predicted value for fish of the same mean length.

RESULT AND DISCUSSION

Length weight relationship

Investigating 300 individuals over three months the total length was found ranging from 11.8 cm to 25.4 cm, standard length of 9.9 cm to 21.4 cm and weighing from 4.36 g to 52.71 g are given in the Table 1. The transformation into linear regression of total length-weight and standard length-weight relationship of the species and their consequent exponential forms are also represented in Table1.

Table 1. Length-weight relationship of *Harpadon nehereus*

Length type	Number of individuals studied	Range of length (cm)	Range of weight (g)	Slope (b)	Power curve equation	Linear form of Power curve equation	Correlation coefficient (r)
Total length	300	11.8-25.4	4.36-52.71	2.404	W= 0.02019 TL $^{2.404}$	ln W= -3.902+ 2.404 TL	0.834
Standard length	300	9.9-21.4		2.169	W= 0.05830 SL $^{2.169}$	Ln W= -2.842+ 2.169 SL	0.802

The total length - weight relationship of non-linear and linear forms were W=0.0202 TL$^{2.404}$ and ln W= -3.0921 + 2.4039 ln TL, respectively with the correlation co-efficient (r) of 0.0.834. Likewise, the standard length-weight relationship of non-linear and linear forms were found to be W=0.0583 SL$^{2.1693}$and ln W = -2.8421 + 2.1693 ln SL, in that order with the correlation co-efficient (r) of 0.803.

The variance of 'b' for both total length and standard length were 0.0090 and 0.0088 respectively. Using the value of 't'i.e. 1.96 from the statistical table the 95% confidence limits were estimated as 2.218 to 2.590 for total length-weight relationship and for standard length-weight relationship, 1.985 to 2.353 . Exploration on length-weight data of H. nehereus showed allometric growth pattern in view of the fact that neither of the two includes '3', the expected value if the relationship between weight and length was cubic and for isometric growth pattern as well.

The present results are in harmony with the earlier reports of Krishnayya (1968), Bapat (1970) and Kurian et al.,(1992) who have also investigated allometric growth pattern in H. nehereus from the Bay of Bengal (b=3.2657) water, Arabian Sea (3.4444 in female and 3.7169 in males) and north west coast (2.0279) respectively. Contrary to the above consequences Nurul Amin (2001) and Bapat et al. (1951) have documented the isometric growth pattern of H. nehereus with 'b' values 3.051 and 2.889, respectively from the Bay of Bengal and Arabian Sea. The 'b' values are known to range between 2.5 to 4 in fishes (Hile, 1936 and Martin, 1949) and in majority of cases the value deviated from 3. According to Mitra (2001) the 'b' values in various species of Hooghly estuary ranged from 2.9615 to 2.3686. Beverton and Holt (1957) reported that adult fishes follow an isometric growth pattern. The inconsistency in 'b' values could be attributed to environmental features, food availability and physiological factors including sex and phase of life (LeCren, 1951; Ricker, 1975).

Table 2. The relative condition factor (CF$_w$) of *Harpadon nehereus* during observation

Month	Mean weight (g) $\overline{W}$	Predicted weight (g) W$_{pred}$	Relative condition factor (CF$_w$)
April	25.61	26.19	0.978
May	29.41	28.48	1.033
June	31.97	29.15	1.097

Relative condition factor (CF$_w$)

The relative condition factor (CF$_w$) of H. nehereus during different month is represented in Table 2. The mean CF$_w$ value of the individuals calculated during three months of observation was found to be 1.036. The monthly fluctuation of the relative condition factor (CF$_w$) showed the lowest in the month of April (0.978) and the highest in the month of June (1.097) though the values exhibited no noteworthy changing pattern during these months as it was 1.033 in May. The relative condition factor is calculated to study the seasonal variation in the condition of fish during different stages of growth and reproduction (King, 2007). They indicate the physiological state and general well-being of fish (Brown, 1957). Krishnayya (1968) recorded the lowest condition factor (1.7392 to 1.849) of H. nehereus in the period between April to June. Moreover, he also noticed the CF$_w$ values of more than 2 from the month of October through March. On the other hand, Bapat (1970) observed an increased condition factor of the same species in the month of April and low condition factor values during December to March. Nurul Amin (2001) found the CF$_w$ values in the range between 0.908 and 1.22. The lower CF$_w$ value in observation period (April to June) as noticed in present study, might be due to the metabolic strain of spawning, spent condition and lower feeding rate of the species since the main spawning season of the species is reported to be December to March (Bapat, 1970). Different studies showed higher CF$_w$ value in the month of February which might be due to peak spawning period and fully fecund gonads. The increased CF$_w$ value after May could also attributed to be the peak feeding period for the species, as observed by Bapat (1951). Such relations of higher CF$_w$ values during peak spawning periods and lower values after spawning periods has also been documented in *Tenualosa ilisha* (Khan et al., 2001). Fluctuations in the CF$_w$ values are common in fishes because of food availability, differential feeding intensity, size of the fish and average reproductive stage of the stock (King, 2007; LeCren, 1951; Thakur, 1975).

CONCLUSION

In this study, the length-weight relationship of *Harpadon nehereus* samples collected from a local market was investigated over three months as part of an ongoing annual research. The findings indicate that the sample of Bombay duck individuals caught by the local gill netters followed an allometric growth pattern. The result of the study also implies that the lower values of relative condition factor was due to sampling immediately after peak spawning season which gradually changes with food availability, seasons and reproductive stages of the stock. During recent years a decreasing trend in Bombay duck fishery is being observed which might be a result of pollution, over exploitation, ineffective management and lack of information on growth in some cases. Studying population dynamics and growth parameters is necessary for proper stock assessment of a fishery where the ultimate goal is to maintain a sustainable yield in future. Present study would help to this extent. In addition further attention should be paid to for improved management of the stock.

COMPETING INTEREST

The author has read and understood RALF policy on declaration of interests and declares that she has no competing interests.

ACKNOWLEDGEMENT

The author wish to expresses her profound gratitude to Professor Dr Zoarder Faruque Ahmed, Department of Fisheries Management, Bangladesh agricultural University and is grateful to the Faculty of Fisheries, Chittagong Veterinary and Animal Sciences University for laboratory facilities and technical support.

REFERENCES

1. Bapat SV, SK Banerji and DV Bal, 1951. Observation on the biology of *Harpadon nenereus* (Ham.). Journal of Zoological Society India, 3: 441 - 356.
2. Bapat SV, 1970. The Bombay duck, *Harpadon nehereus* (Ham.). Bulletin of Central Marine Fish Research Institute, 21: 1 - 66.
3. Beverton RJH and SJ Holt, 1957. On the dynamics of exploited fish populations. Fishery investigations / Ministry of Agriculture, Fisheries and Food, Great Britain (2 sea fish), 19: 533.
4. Brown MR, 1957. Experimental studies on growth. In the physiology of Fishes. Vol-1 (Ed.Brown, M.E.) New York, Academic Press. 361 – 400.
5. Gopal B and M Chauhan, 2006. Biodiversity and Its Conservation in the Sundarban Mangrove Ecosystem. Auatic Sciences 68:338-354.
6. Hile R, 1936. Age and growth of Cisco, *Leucichthyes artedi* (Le Surr) in the lakes of three northern highland, Wisconsin. Wise Bulletin of United States Business Fisheries, 48: 211-317.
7. Hoq ME, MA Wahab and MN Islam, 2006. Hydrographic status of Sundarbans mangrove, Bangladesh with special reference to post-larvae and juveniles fish and shrimp abundance. Wetland Ecology and Management, 14: 79-93.
8. Kamal M, BC Biswas, L Yasmin, KM Azimuddin and MN Islam, 2001. Influence of temperature on the gel-forming ability of some under-utilized marine fish species in Bangladesh. Pakistan Journal of Biological Science, 4: 1525-1542.
9. Khan MA, D Kumar and R Sinha, 2001. On some biological aspects of *Tenulosa ilisha* (Hamilton-Buchanan) from Hooghly estuary. Journal of Inland Fisheries Society of India, 33: 38 - 44.
10. King, M, 2007. Fisheries Biology, Assessment and Management, Second Edition, Blackwell Publishing Ltd.,Oxford, UK, pp: 172-238.
11. Krishnayya CH, 1968. Age and growth of *Harpadon nehereus* (Ham.) and its fishery in the Hooghly. Journal of Zoological Society of India, 20: 129 - 147.
12. Kurian A and KN Kurup, 1992. Stock assessment of Bombay duck *Harpadon nehereus* (Ham.) off Mahasashtra coast. Indian Journal of Fisheries, 39: 243-248.

13. Larkin PA, 1977. An epitaph to the concept of maximum sustainable yield. Transactions of the American Fisheries Society, 106:1-11.

14. LeCren ED, 1951. The length-weight relationship and seasonal cycle in gonad weight and condition of the pearch, Pereafluviatilis. Journal of Animal Ecology, 20: 210 - 219.

15. Martin WR, 1949. The mechanisms of environmental control of body form in fishes. University of Toronto Studied in Biology, 58:1 - 91.

16. Mitra PM, 2001. Length-Weight relationship of some commercially important fish species of Hooghly estuary. Journal of Inland Fisheries Society of India, 33: 12 - 14.

17. Nurul Amin S M, 2001. Study on age and growth VPA analysis and relative condition factor of *Horpadon nehereus* (Ham.) from the neretie water of Bangladesh. Online Journal of Biological Science, 1: 192 - 194.

18. Rahman AKA, 1989. Freshwater fishes of Bangladesh. Zoological Society of Bangladesh. Department of Zoology, University of Dhaka, 364.

19. Rahman AKA, 2005. Freshwater fishes of Bangladesh, second edition. Zoological Society of Bangladesh, University of Dhaka, Dhaka, Bangladesh, 394.

20. Ricker WE, 1975. Computation and interpretation and biological statistics of fish populations. Bulletin of Fisheries Research Board of Canada, 191: 382.

21. Rupshankar C, 2010. Improvement of cooking quality and gel formation capacity of Bombay duck (*Harpadon nehereus*) fish meat. Journal of Food Science and Technology, 47: 534-540.

22. Talwar PK and AG Jhingran, 1991. Inland fishes of India and adjacent countries, volume 2, Oxford IBH Publishing Co Pvt Ltd, New Delhi-Calcutta, pp. 724-725.

23. Thakur NK, 1975. On the Length – weight relationship and relative condition factor in *Clarias batrachus* (Linn.). Proceedings of the National Institute of Science, India, 45: 197 – 202.

12

FISH FARMING STATUS AT SREEMANGAL UPAZILA OF MOULVIBAZAR DISTRICT, BANGLADESH

Baadruzzoha Sarker[1] and Muhammad Forhad Ali*[2]

[1]Production Officer, BRAC Fish Hatchery, Sreemangal, Moulvibazar-3214, Bangladesh and [2]Department of Aquaculture, Sheikh Fajilatunnesa Mujib Fisheries College, Melandah, Jamalpur, Bangladesh; *PhD Fellow, Department of Fisheries Biology and Genetics, Bangladesh Agricultural University, Mymensingh-2202, Bangladesh

*Corresponding author: Muhammad Forhad Ali; E-mail: mforhad.fc@gmail.com

ARTICLE INFO	ABSTRACT
Key words Fish farming, Stocking density, Feed, Fish production, Sreemangal	The potential and prospect of fish farming of Sreemangal, Moulvibazar was carried out from January to December 2015 by interviewing of 90 fish farmers with a well-structured questionnaire. The survey revealed that 40% of the ponds were seasonal and 60% perennial, of which 85% with single ownership and 15% accompanied by multiple ownership. The average pond size was 0.13 ha and depth 2.6 m. The ponds were prepared using of lime, cow dung, urea and TSP at the rate of 250, 250, 40 and 20 kg/ha, respectively. Nearly all of the farmers practiced polyculture with Indian major carps and exotic carps. Fish was stocked from March to June and average stocking density was 15,500 fingerlings/ha. To sustain natural food production, farmers generally used cow dung, urea and TSP at the rate of 2,600, 300 and 150 kg/ha/yr, respectively. Healthy environmental condition was maintained by applying lime and salt at 600 and 60 kg/ha/yr, respectively. The fish were fed by supplementary feed (45% farm) e.g., rice bran and mustard oil-cake with an average quantity of 2,200 and 550 kg/ha/yr, respectively and artificial pellet feed (55% farm). The average fish production, production cost and profit were found 2,945 kg/ha/yr, 1,25,940 and 94,935 Tk./ha/yr, respectively. The major constraints for sustainable pond fish farming were non-availability of fish fingerlings during stocking period, insufficient water in dry season, high production cost, poor technical knowledge, lack of money etc. By establishing more hatcheries, arranging training at farm level, providing interest free or at lower interest loan to the farmers the existing fish production could be increased.

INTRODUCTION

Aquaculture and fisheries at the moment is one of the most important potential sectors of the national economy, accounting to 3.69% of national GDP, 22.60% of agricultural GDP and 2.09% of foreign export earnings (FRSS, 2015). In Bangladesh no other sector depicted progress prospective more visibly than fisheries. The total fish production in Bangladesh in the fiscal year 2013-14 was estimated as 3.55 million tons, of which 1.96 million tons (55.15%) were obtained from inland aquaculture, 0.99 million tons (28.07%) from capture fisheries and 0.60 million tons (16.78%) from marine fisheries (FRSS, 2015). Among the global food production systems, aquaculture is widely professed as a key weapon in the global struggle against poverty and hunger. Aquaculture production, particularly pond aquaculture may be are liable source of attaining increased fish production so as to provide and feed the continually rising population of the planet (FAO, 2010). Fish and fishery resources play a vital role in improving the socio-economic condition, combating malnutrition, earning foreign currency and creating employment opportunities in Bangladesh (Bhuiyan et al., 2011). More than 17.5 million people are engaged with this sector on full time and part time basis (FRSS, 2015).

In the country, around 46% of children between the ages 6 and 7 years are stunted and 70% are exhausted due to malnutrition (Ahmed et al., 2007). The greater importance should be given to meet the animal protein deficit among the people as well as to augment fish production in this country through proper management of open water fishery and aquaculture. However fish production from open water bodies is declining progressively (DoF, 2012). In the past, the natural ecosystems supported huge and diverse biodiversity. Currently, volume of the most natural water bodies were reduced due to siltation, construction of flood control dam and concurrently polluted by agricultural, industrial and metropolitan waste as a contaminant and those were accumulated by runoff into natural water bodies. Accordingly, aquatic organisms were silently suffered to the sub-lethal toxicity by different types of chemicals, heavy metals and pollutants (Bernet et al., 1997). Consequently, capture fish production is also falling down due to manmade interventions and natural disasters. On the other hand, it is hopeful that aquaculture practice has become a promising and gainful methodology to attain self-sufficiency in food sector and also to alleviate poverty in developing country like Bangladesh (Ahmed, 2003). Additionally, more returns also come from the pond aquaculture (DoF, 2012). Aside from steadfast self-employment opportunities from fish farming, pond fish farming offers various livelihood prospects for operators and employees of hatcheries and seed nurseries and for seed traders and other mediators. Pond fish farming has also been proved to be an advantageous business than rice cultivation. Thus many farmers in rustic areas are converting their rice field into aquaculture pond. A lot of people in rural areas have taken fish farming activities as their secondary profession and most of them occupied in pond fish farming to enhance their socio-economic status (Ahmed, 2003).

Sreemangal is an upazila of Maulvibazar District in the Division of Sylhet, located at north east part (24.3083°N 91.7333°E) in Bangladesh. It has a total area of 450.73 km², with a population of 3, 18,025 (BBS, 2011). Sreemangal is the business nucleus of the district of Moulvibazar. Her economy has made of world-class tea gardens and tourism. Presently fisheries sector is emerging here and a considerable number of people have been engaged in fish culture practices in this hilly riverine area. For the development of cultural system and future planning, the information concerning present aquaculture practices at the grass root level is an important criterion. This study observed the present fish farming status in Sreemangal upazila and sought any constraints associated with the culture systems.

MATERIALS AND METHODS

Study Area and Target Farmers

The study was conducted at Sreemangal upazila in Moulvibazar district for obtaining detailed information about pond fishery resources and constraints associated. The data were collected fortnightly from 90 randomly selected fish farmers of 9 unions covering the selected study area from January 2015 to December 2015.

Data Collection Methods

The necessary data were collected using questionnaire and crosscheck interviews. The draft questionnaire was tested by the opinion of 10 fish farmers and much awareness was given to any new information which was not designed to ask but was considerable and informative to accomplish the objectives. The questionnaire was changed, modified and rearranged according to the experiences. Data were collected from fish farmers to get comprehensive information on their fish farming systems and various obstacles concerned with it. For collecting data both individual and group interviews were conducted at the pond sites and or in the house of the farmers. In favour of this study one of the PRA (Participatory Rural Appraisal) tool and FGD (Focus Group Discussion) was conducted to obtain more accurate data (Chambers, 1992; Nabasa et al., 1995). A total of 10 FGD sessions were conducted where each group size of FGD was 8 to10 farmers. The FGD session was held in front of village shops, under the big trees, farmer's house and school premises. At the beginning of the interview, a brief introduction about the objectives of the survey was given to each of the farmers/FGD groups and assured them that all information would be kept confidential. Each question was explained clearly and asked systematically for their sound understanding. After collecting the data through questionnaire interviews and FGD, crosscheck interviews were conducted with Upazila Fisheries Officer, Assistant Fisheries Officer, relevant NGO workers, Chairman and Members of the Union Councils.

Data Processing and Analysis

The collected data were scrutinized and summarized carefully before the actual tabulation. Some of the data e. g., pond area and pond depth were collected into local units and converted into data with standard units. Then the data were tabulated into a preliminary data sheet of a computer and compared with computer spreadsheets to ensure the accuracy of the data entered. After data entry, the data were analyzed through statistical method using Microsoft excel.

RESULTS AND DISCUSSION

Pond Category and Ownership

In the study area, ponds were found of two categories-homestead and commercial. The homestead and commercial ponds were 75% and 25%, respectively. Between the two categories of ponds 40% were seasonal and 60% perennial. The water level of perennial ponds declined in dry season, then some farmers used pump water in their ponds. Seasonal ponds were found unsuitable for fish culture throughout the dry season. The pond types were found 46% seasonal and 54% perennial in Rajshahi district (Ali et al., 2008), 37% ponds were seasonal and 63% perennial in Tangail sadar upazila (Saha, 2004). It was observed that the uppermost number of ponds (85%) was occupied by the single owners followed by multiple owners (15%). Whereas, about 54% of the total pond were in single ownership, 34% were belongs to joint ownership and the rest 12% ponds were under public or organization property in Demra, Dhaka (Quddus et al., 2000), and 52% ponds were found under single ownership, 21% in multiple ownership and remaining 27% as leased pond in Tangail sadar upazila (Saha, 2004). The multiple pond ownership was a main constraint for pond aquaculture (Ali and Rahman, 1986, Mollah et al., 1990 and Hossain et al., 2002).

Pond Size and Depth

Pond size is a vital feature as every management events are intended allowing for the size of ponds. It was observed that the average pond size was 0.13 ha with a range from 0.05 ha to 0.81 ha in the study area. As well as, the average pond size was found as 0.12 ha in Gazipur (Rahman, 2003), 0.19 ha in Tangail sadar (Saha, 2004), 0.21 ha in Dinajpur sadar (Saha, 2003) and 0.22 ha in Trishal, Mymensingh (Sheheli et al., 2013). Fish culture efficiency diverse with the size of ponds (Khan, 1986). The average depth of pond in the study area was found 2.6 meter. However in Bangladesh it varied from 2 to 5 meter (DoF, 2010) which corresponds well with the present study.

Culture Season and System

In the study site, the duration of fish farming was from June to November in case of seasonal ponds and March to November in perennial ponds. Fish fry were stocked when they become available in March to June. The peak period of carp polyculture was observed from March to December (Rahman, 2003) and also commencing from April to December (Ahmed, 2003). Furthermore, two culture seasons were practiced at Fazilpur and Sunderban union under Dinajpur sadar upazila; one was from July to December and the another was from February to June (Saha, 2003). The majority of farmers (98%) adopted polyculture and only 2% ponds were under integrated culture system. The idea of polyculture was based on utilizing different niches by various fish species. Therefore, a more complete use was made of the food resources and space available in polyculture than in monoculture (Anil et al., 2010).

Pre-stocking Managements

Pond preparation is one of the important tasks to obtain more production in fish farming. Several steps were followed by the farmers before stocking of fish. These are dike repairing, removal of the excessive mud from pond bottom; aquatic weed control, eliminate predatory and undesirable species, lime and fertilizer application, etc. In pre-stocking management, about 95% of the farmers controlled aquatic weeds manually. For eradicating undesirable species, most of farmers (90%) used netting method. Some of them (10%) used rotenone and phostoxin but, did not follow any recommended dose. Almost all farmers dried their ponds after harvesting of fish in the dry season, and they used lime at 250 kg/ha. In Trishal, Mymensingh about 86% farmer dried their pond after deteriorating water quality and, among them 54% and 46% owner applied lime at 247 and 370 kg/ha, respectively during pond preparation (Sheheli et al., 2013). In the study area, farmers used fertilizers mainly in the form of cow dung @ 250 kg/ha, urea @ 40 kg/ha and TSP (triple super phosphate) at 20 kg/ha. The purpose of using fertilizers in the ponds was to increase the production of natural food (phytoplankton, zooplankton and benthic organisms), in that way to augment fish production. In the study area, the use of cow dung was widespread due to being fairly cheap and available.

Cultured Species and Stocking Density

In polyculture system, farmer cultured mainly Indian major carp such as rohu (*Labeo rohita*), catla (*Gibelion catla*), mrigal (*Cirrhinus cirrhosus*), kalibaush (*Labeo calbasu*), and exotic carps like, silver carp (*Hypophthalmichthys molitrix*), grass carp (*Ctenopharyngodon idella*), punti (*Puntius gonionotus*) and common carp (*Cyprinus carpio var communis*), and other alien species viz., tilapia (*Oreochromis niloticus*) and pangus (*Pangasius sutchi*). It was found that all farmer stocked hatchery produced fry. The average stocking density was determined as 15,500 fingerlings/ha. The average stocking density was recorded as 14,675 fry/ha at Shahrasti upazila in Chandpur (Pravakar et al., 2013), 17,262 fry/ha at Mithapuqur upazila, Rangpur (Alam, 2006) and 25,250 fry/ha/yr in Gazipur (Rahman, 2003). The stocking density of carp was maintained 10,621-13,091 fry/ha (DoF, 2005). However, it was observed a range of stocking density from 10,000-31,000 fry/ha at a village of Mymensingh district (Hossain et al., 1992).

Farm and Fish Health Managements

To maintain natural food production, it is essential to apply both organic and inorganic fertilizer into pond throughout the culture period. It was observed that majority of the farmers used cow dung and only a few farmers used poultry droppings as organic fertilizer. The fish farmers generally used cow dung, urea and TSP at the rate of 2,600, 300 and 150 kg/ha/yr, respectively on a regular basis, or four to five times during the culture period. Also, the average dose of organic fertilizer (cow dung) was 2,330 kg/ha/yr and inorganic fertilizer such as, urea at 387 kg/ha/yr and TSP at 176 kg/ha/yr were used in Tangail sadar upazila (Saha, 2004) and the doses of same organic and inorganic fertilizers were 11,075 and 739 kg/ha/yr, respectively (Rahman et al., 1998). All the farmers frequently used lime to maintain suitable water quality and reduce deleterious gases on an average rate at 600 kg/ha/yr of the study area. To get more fish productions farmer used lime at the rate of 247 kg/ha/yr in Panchagar (Islam and Haque, 2010). In the present study, 70% farmers applied salt three to four times at the commencement and during the winter season at 60 kg/ha/yr to keep healthy environmental condition and avoid infections. Ten percent farmers used some of the chemicals like copper sulphate, $KMnO_4$, dipterex, methylene blue, malachite green and calcium hypochlorite to control aquatic weeds, pests, parasites and detrimental species and prevent different types of diseases. Cent percent farmer used more or less similar chemicals and other medicinal substances for controlling aquatic weeds, pest, predators and harmful species in their ponds in Trishal, Mymensingh (Sheheli et al., 2013). In this study, it was observed that out of 90 farmers, 20% farmers used antibiotics for control of different diseases. It also appeared that 22% farmers used antibiotics (Sheheli et al., 2013). Subsequent to good health management just 25% farmers did not found fish diseases, 65% farmers reported that fish was occasionally affected by diseases, while 10% farmers found disease outbreak every year. The most common diseases were tail and fin rot, epizootic ulcerative syndrome (EUS), argulosis, saprolegniasis, edwarsielosis, and nutritional deficiency. There were 15% fish farmers who did not find fish diseases in the pond, 73% farmer reported that their cultured fish was sporadically affected by diseases, while 12% fish farmers found disease outbreak every year in Trishal, Mymensingh (Sheheli et al., 2013).

Feed and Feeding Practices

Different types of feeds were used in fish production. Farmers mostly used three types of feed such as loose, pellet and green grass. From the survey, it was recorded that 45% of the farmers applied farm made supplementary feed prepared with rice bran and mustard oil-cake (80% and 20% ratio) and 55% farmers used artificial pellet feed. Whereas, approximately 95% farmer used rice bran and mustard oil cake as supplementary feed in Shahrasti, Chandpur (Pravakar et al.,2013), and 80% farmer applied the similar feed in Mithanpukur, Rangpur (Alam, 2006). In the present study, the farmer usually used rice bran and mustard oil-cake to feed the fish at 2,200 and 550 kg/ha/yr, respectively. However, the supplementation of rice bran and oil cake was at 2,730 and 580 kg/ha/yr in Gazipur (Rahman, 2003), 1,250 and 1,212 kg/ha/yr in Rajshahi (Hassanuzzaman, 1997) and, 1,920 and 100-110 kg/ha/yr in Debigonj and Boda upazilas, Panchagar (Islam and Haque, 2010), respectively. In the investigated area, some farmers used green grass and duck weed to feed the fish at undetermined quantity. Typically farmer collected floating and sinking pellet feed from the local market. In the study area it was found that nearly 55% farmers provided feed twice a day, while 25% of farmers supplied thrice a day and 20% applied feed once daily. There were 75% farmers applied feed twice in a day, 20% of farmers applied thrice and only 5% applied feed once per day in Trishal, Mymensingh (Sheheli et al., 2013). The recommended feeding frequency was two or three times per day (DoF, 2009). Both artificial and supplementary feed was used by the farmers, which showed a good sign about the perception of fish culture by the farmers.

Harvesting and Marketing

Though fish were harvested all over the year but the peak harvesting season lied from December to February. In this season around 65% fishes was harvested and remaining 35% was harvested during other season. The similar peak harvesting season was from December to March in Mymensingh (Ahmed, 2003). Also, the peak harvesting season was observed from November to January in Tangail sadar upazila (Saha, 2004) and from October to January in Gazipur (Rahman, 2003). Farmers harvested their fish using cast net and seine net locally known as *berjal*. Farmers were widely used the same nets to harvest fish in Mymensingh (Ahmed, 2003), in Debigonj and Boda upazilas, Panchagar (Islam and Haque, 2010) and in Shahrasti upazila, Chandpur (Pravakar et al., 2013). Harvested fish were kept in aluminum containers or bamboo baskets. From the survey it was found that around 95% of the fishes were sold by the farmers to the local agent and the rest 5% consumed by the households and given to the relatives. Although, nearly 72.5% farmers sold their captured fish to the fish traders, while 17.5% of them sold the fish to the retailers (*Foria*) and the rest 10% of them to the fish agent (Islam and Haque, 2010) and around 80% fish were sold by the farmers to local *paikers* and the rest 20% consumed by the households and given to the relatives in Shahrasti upazila, Chandpur (Pravakar et al., 2013).

Fish Production, Cost and Profit

In Sreemangal upazila the average yield of fish was found 2,945 kg/ha/yr. Likewise, the average production was 2,900 kg/ha/yr in Shahrasti, Chandpur (Pravakar et al., 2013), 2,925 kg/ha/yr in Gazipur district (Rahman, 2003) and also the same carp production was found in Bhaluka, Mymensingh (Kamruzzaman, 2011) and 2,940 kg/ha/yr in Moulvibazar (FRSS, 2015). On the other hand, mean (±SD) fish production was found 3,598.72±785.83 kg/ha/yr in carp polyculture of Rajshahi and Natore districts (Mohsin et al., 2012) and 3,743 kg/ha/yr in Mymensingh (Biswas, 2003). The annual production varies because of differences in pond size, depth, categories, feed, seed, other inputs and management practices. In the study area it was found that average total cost of fish production was 1,25,940 Tk./ha/yr. The mean total cost of fish production in Shahrasti, Chandpur was observed as 80,850 Tk./ha/yr (Pravakar et al., 2013). The production cost of fish was higher due to the increase of the price of fingerlings, feed, fertilizers, drugs, chemicals and labour. Maximum amount of production cost was spend for fish feed (31.80%) followed by labour (24.22%), fingerlings (20.49%), water pumping and electricity (5.34%), fertilizers (4.39%), lime (3.94%), fish marketing (2.54%), miscellaneous (2.38%), fish harvesting (2.06%), cow-dung/organic manure (1.59%), drugs/chemicals (0.79%) and salt (0.40%). Similar production cost was assessed in Shahrasti, Chandpur spend for fish feed (28%) followed by fingerlings (25%), water pumping and electricity (9%), labour (8%), lime (8%), fish marketing (7%), fertilizers (5%), miscellaneous (4%), fish harvesting (3%), cow-dung/organic manure (2%) and drugs/chemicals (1%) (Pravakar et al., 2013). The average profit from the study area was found 94,935 Tk./ha/yr. In case of extensive, improve extensive and semi-intensive categories of culture net profit was observed from fish culture were 46,600, 63,000 and 92,000 Tk/ha/yr, respectively in Demra, Dhaka (Quddus, et al., 2000). The mean profit from fish culture of Shahrasti, Chandpur was found as 1,19,400 Tk./ha/yr. (Pravakar et al., 2013). The profit in fish culture was found fairly similar to the previous study.

Constraints of Fish Farming

The interviewed fish farmers faced a diversity of difficulties and constraints that affected the fish farming activities as well as their livelihood. On the basis of the survey 30% farmers confronted by non-availability of fish fry during stocking period as the single most important problem for fish farming. The other problems raised by the farmers were insufficient water in dry season (23%), high production cost (16%), poor technical knowledge (15%), lack of credit source (8%), lack of money (4%), low price of the produced fish (4%). Beside these, some fish ponds were occasionally inundated by flash flood in the study area. In other survey in Shahrasti, Chandpur it was exposed that 30% of the fish farmers identified fish disease as the only most momentous trouble in fish farming, following non availability of fish fry (20%), insufficient water in dry season (16%), pouching (14%), poor technical knowledge (10%), lack of money (6%) and lack of quality feed (4%) (Pravakar et al., 2013). In addition, it was reported that lack of scientific knowledge and suitable technology, less extension services on aquaculture training, turbidity, non availability of quality fish seeds at proper time, occurrences of fish diseases, water scarcity during drought season, pond water irrigation for crop fields were found in Mohanpur, Rajshahi (Zaman et al., 2006).

RECOMMENDATIONS

According to the outcomes of the present study, the subsequent recommendations were made for sustainable pond fish farming of farmers in Sreemangal upazila under Moulvibazar district. The problem of multiple ownerships can be solved by leasing the pond to a person interested in fish culture or through cage or pen culture by different owners. Supply of net and other harvesting and marketing equipments to the farmers with less fare may reduce harvesting and marketing cost. Government and other organization should play their assigned task by disseminating technology to the farmers and arranging essential training for scientific methods of aquaculture. Such training will aid to identify and solve the problems related to the fish farming. To supply quality fish seed to the farmers more hatcheries should be established by the help of government and NGO. Money lending from bank is lengthy and bureaucratic process and also a question of bride. The accessibility of credit at a low interest rate from bank should be eased and existing problems should be removed. Government should take necessary measures for proper extension work, which will improve the social, moral and scientific education among the farmers and neighbors; therefore the fish production will ultimately be increased.

CONCLUSION

The study showed that fish production was satisfactory and all farmers made a profit from fish farming. While the potential benefits are great, lack of aquaculture friendly credit support and inadequate technical assistance were constraints to the sustainability of fish farming. It is therefore, necessary to provide institutional, organizational, and government support for sustainable fish farming. Additionally, a cost effective research-extension-farmers association technique is a prime need to increase fish production.

REFERENCES

1. Ahmed F, 2003. Comparative study on carp polyculture of three different NGOs in Mymensingh district. MS Thesis, Department of Aquaculture. Bangladesh Agricultural University, Mymensingh. pp. 65.

2. Ahmed MNU, 2003. Fisheries sector in Bangladesh. Economy and Development of livelihood. Fish Fortnight Compendium, Department of Fisheries, Bangladesh, pp. 86.

3. Ahmed N, MA Wahab and SH Thilsted, 2007. Integrated aquaculture-agriculture systems in Bangladesh: Potential for sustainable livelihoods and nutritional security of the rural poor. Aquaculture Asia, 12: 14-22.

4. Alam G, 2006. Status of fish farming and livelihoods of fish farmers in some selected areas of Mithapuqur upazila in Rangpur district. MS Thesis, Department of Fisheries Management, Bangladesh Agriculture University, Mymensingh. pp. 59.

5. Ali MH and MI Rahman, 1986. An investigation on some socioeconomic and technical problems in pond fish culture in two districts of Bangladesh. Bangladesh Journal of Aquaculture, 8: 47-51.

6. Ali MH, MD Hossain, ANGM Hasan and MA Bashar, 2008. Assessment of the livelihood status of the fish farmers in some selected areas of Bagmara upazilla under Rajshahi district, Journal of Bangladesh Agricultural University, 6: 367-374.

7. Anil SK, B Gunalan, KL Jetani, GK Trivedi and P Soundarapandian, 2010. Determine the Economic Feasibility of the Polyculture System (Giant Tiger Shrimp and Mullet). African Journal of Basic and Applied Sciences, 2: 124-127.

8. BBS (Bangladesh Bureau of Statistics), 2011. Population and Housing Census 2011. Bangladesh Bureau of Statistics, Statistics and Informatics Division, Ministry of Planning.

9. Bernet D, H Schmidt, W Meier, P Burkhardt-Hol and T Wahli, 1997. Histopathology in fish, Proposal for a protocol to assess aquatic pollutions, Journal of Fish Diseases, 22: 25-34.

10. Bhuiyan AS, S Akther and N Aktar, 2011. Present status and fish seed production of the hatcheries of six upazilas of Rajshahi District. University Journal of Zoology Rajshahi University, 30: 29-32.

11. Biswas D, 2003. Study of the impacts of Aquaculture in and around fish farms in Mymensingh district. Progressive Agriculture, 11: 243-249.

12. Chambers R, 1992. Rural Appraisal: Rapid, Relaxed and Participatory, IDS Discussion Paper No. 311, Institute of Development Studies (IDS), Brighton, UK.

13. DoF (Department of Fisheries), 2010. Fisheries Fortnight Compendium, Department of Fisheries, Ministry of Fisheries and livestock, Dhaka, Bangladesh.

14. DoF, 2005. Basic Training of Fish Culture, Department of Fisheries, Dhaka, Bangladesh. pp. 87.

15. DoF, 2009. Department of Fisheries, Ministry of Fisheries and Livestock, Jatiyo Matshya Saptaho, pp. 54-56.

16. DoF, 2012. Fishery Statistical Yearbook of Bangladesh. Fisheries Resources Survey System, Department of Fisheries, Dhaka, Bangladesh.

17. FAO, 2010. State of World Aquacultutre-2010. Fisheries Department. FAO Fisheries Technical Paper, 500: 21-26.

18. FRSS, 2015. Fisheries Statistical Report of Bangladesh. Fisheries Resources Survey System (FRSS), Department of Fisheries, Bangladesh. 31: 1-57.

19. Hassanuzzaman AKM, 1997. Comparative study on pond fish production under different management systems in some selected areas in Rajshahi district. M. S. Thesis. Department of Agricultural Production Economics. Bangladesh Agricultural University, Mymensingh. pp. 76.

20. Hossain MS, S Dewan, MS Islam and SMA Hossian, 1992. Survey of pond fishery resources in a village of Mymensingh district. Bangladesh Journal of Aquaculture, 14-16: 33-37.

21. Hossain MA, MAR Khan and MA Mannan, 2002. Present status of pond fishery in Dhamaihat upazila, Naogoan. University Journal of Zoology Rajshahi University, 21: 79-80.

22. Islam MR and MR Haque, 2010. Impacts of Northwest Fisheries Extension Project (NFEP) on pond fish farming in improving livelihood approach. Journal of Bangladesh Agricultural University, 8: 305–311.

23. Kamruzzaman M, 2011. Study of aquaculture practices of Bhaluka Upazila, Mymensingh. MS Thesis, Department of Fisheries Management, Bangladesh Agricultural University, Mymensingh.

24. Khan MS, 1986. Socio-economic factors in the development of fisheries. Bangladesh Journal of Agricultural Economics, 10: 43-47.

25. Mohsin ABM, MN Islam, MA Hossain and SM Galib, 2012. Constraints and prospects of carp production in Rajshahi and Natore districts, Bangladesh. University Journal of Zoology Rajshahi University, 31: 69-72.

26. Mollah AR, NSI Chowdhury and MAB Habib, 1990. Input output relation in fish production under various pond sizes, ownership patterns and constraints. Bangladesh Journal of Training and Development, 3: 87-101.

27. Nabasa J, G Rutwara, F Walker and C Were, 1995. Participatory Rural Appraisal: Practical Experience. Natural Resources Institute (NRI), Greenwich University, London.

28. Pravakar P, BS Sarker, M Rahman and MB Hossain, 2013. Present Status of Fish Farming and Livelihood of Fish Farmers in Shahrasti Upazila of Chandpur District, Bangladesh. American-Eurasian Journal of Agricultural & Environmental Sciences, 13: 391-397.

29. Quddus MA, MS Rahman and M Moniruzzaman, 2000. Socio-economic conditions of the pond owners of Demra, Dhaka. Bangladesh Journalof Fisheries Research, 4: 203-207.

30. Rahman MA, MZ Sofiquzzoha and M Nurullah, 1998. Efficiency of pond fish production in Bangladesh. Bangladesh Journal of Agricultural Sciences, 25: 235-239.

31. Rahman MM, 2003. Socio-economic aspects of carp culture development in Gazipur, Bangladesh. M. S. Thesis, Department of Agricultural Economic. Bangladesh Agricultural University, Mymensingh. pp. 83.

32. Saha MK, 2003. A study on Fish Production Technology in North-West Bangladesh. MS Thesis, Department of Aquaculture. Bangladesh Agricultural University, Mymensingh. pp. 71.

33. Saha SK, 2004. Socio-economic aspects of aquaculture in Tangail Sadar Upazila, MS Thesis, Department of Aquaculture. Bangladesh Agricultural University, Mymensingh. pp. 77.

34. Sheheli S, K Fatema and SM Haque, 2013. Existing Status and Practices of Fish Farming in Trishal Upazila of Mymensingh District. Progressive Agriculture, 24: 191-201.

35. Zaman T, MAS Jewel and AS Bhuiyan, 2006. Present status of pond fishery resources and livelihood of the fish farmers of Mohanpur Upazila in Rajshahi District. University Journal of Zoology Rajshahi University, 25: 31-35.

13

EFFECT OF SALT AND SMOKE ON QUALITY AND SHELF LIFE OF SALT-SMOKE-DRIED BATASHI (*Neotropius atherinoides*) KEPT AT DIFFERENT STORAGE CONDITION

Md. Masud Rana[1] and Subhash Chandra Chakraborty[2*]

Department of Fisheries Technology, Faculty of Fisheries, Bangladesh Agricultural University, Mymensingh-2202, Bangladesh

*Corresponding author: Subhash Chandra Chakraborty, E-mail: subhash55chakraborty@yahoo.co.uk

---

ARTICLE INFO

Key words

Batashi,
Salt-smoke-dried,
Shelf-life,
Chaluni

ABSTRACT

This study was carried out to assess the changes in microbiological and biochemical aspects of fresh, smoked, salt-smoke-dried (SSD), control dried (CD; treated without salt and smoke) batashi (*Neotropius atherinoides*) during storage at ambient temperature (26-28°C) and refrigeration temperature (4°C). The main objective of this study was to investigate the effect of natural preservatives such as salt and smoke on the shelf life of the product over the storage period. The moisture content of fresh batashi was 75.81% whereas 18.10% for SSD batashi. The initial value of protein, fat and ash content of SSD batashi was 65.93, 17.09 and 16.90% on dry matter basis. During storage period (60 days), the percentage of moisture increased whereas protein, fat, and ash content considerably decreased. After two month storage at ambient temperature the protein, lipid and ash content for SSD batashi were 65.0, 15.9 and 15.95%, respectively whereas the values of the same parameters stored at refrigeration temperature were 65.03, 16.16 and 16.12%, respectively on dry matter basis. The initial TVB-N and SPC value for SSD batashi was 6.10mg/100g and 1.14×10^4 CFU/g respectively. During storage period TVB-N and SPC value slowly increased and the values reached to 17.94 mg/100g and 4.2×10^4 CFU/g, respectively for salt-smoke-dried batashi, stored at ambient condition whereas 11.20 mg/100g and 2.42×10^4 CFU/g, respectively for the products stored at refrigeration temperature. Because of using salt and smoke as natural preservatives, no yeast or mould was detected in salt-smoke-dried batashi fish. Therefore, it can be inferred that salt and smoke treated dried fish products has longer shelf life than plain dried product (without treated with salt and smoke). From the overall performance, the products kept at refrigeration temperature showed better shelf life than the products stored at ambient temperature.

INTRODUCTION

Fish provides a high source of protein required in the diets of man as it contains essential nutrients such as vitamins, fats and minerals which help in the maintenance of life (Ashano and Ajayi, 2003). From the age old system curing of fish is being used as means of preservation for increasing its shelf life and its various use by the consumers at different levels. Curing generally include the methods like, salting, smoking and drying fish are in principle the reduction of moisture to decrease the water activity (a_W) in fish muscle. During post-harvest period large amount of fish are spoiled and wasted due to lack of proper measure for processing and preservation because of the fact that neither we can consume all the fishes caught nor can we transport to other places wherever necessary due to our insufficient handling and transportation system. In other words, proper handling, processing and preservation during post-harvest period are a prerequisite for minimizing the spoilage loss (Clucas, 1981). The processing and preservation of fresh fish is most important since fish is highly susceptible to deterioration immediately after harvest and also to prevent economic losses. Fish Smoking is one of the traditional fish processing methods aimed at preventing or reducing postharvest losses. Smoking involves heat application to remove moisture and it inhibits bacterial and enzymatic actions of fish (Kumolu-Johnson *et al.*, 2009) enhance flavor and increase utilization of the fish (Nahid *et al.*, 2016). There are many factors e.g. brine concentration, smoking time, smoke temperature, types of fish species, types of smoke source etc. responsible for the shelf-life and quality of the smoked fish (Salim *et al.*, 2007). In Bangladesh, generally all these curing methods for fish processing mentioned above are well accepted and more or less popular as separate by the consumers at different levels and even two methods altogether for one product is being popular as in case of ready to eat product from hot smoking (where brining and smoking is done together) and salt drying (where salting and drying is done together) (Mansur *et al.*, 1998). It is assumed that the three curing process viz. salting, smoking and drying in combination would produce a new, better quality product probably with a longer shelf-life by significantly reducing water activity in the flesh of fish with characteristic flavor and taste to be preferred by the consumers.

There has been very limited reported work on fish smoking in Bangladesh. In Bangladesh smoked fish is recent addition to the fishery products. Preservation of Small Indigenous Species (SIS) fish is comparatively a new trend or new kind of research activities in this country. Due to high palatable, taste and rich in nutrients freshwater SIS such as batashi (*Neotropius atherinoides*) have been selected for the present research work with an aim to produce salt-smoke-dried fish product to serve as one of the better-accepted quality fishery products showing a longer shelf life when kept at ambient and refrigeration temperature.

MATERIALS AND METHODS

Sample Collection
For the preparation of salt-smoke-dried products from freshwater fish species, batashi; (*Neotropius atherinoides*) was collected as fresh from the Machua bazar, Mymensingh town by direct contact with supplier early in the morning. The collected fresh fish samples were carried in ice stored condition to the laboratory of Fisheries Technology, Bangladesh Agricultural University.

Washing and Dressing of Fish
At first the fishes were washed in potable water, weighed the whole fish on a sensitive balance then dressed (gutting, finning and spinning) and weighed the dressed fish in the laboratory.

Brining, dewatering, air drying and pre-smoking treatment
After draining out water the fishes were allowed for salt treatment. The fishes were immersed into a plastic bucket containing 25% salt solution for 5 min. after that kept the fishes on a plastic tray for drying at room temperature for about 10 minutes. After air drying the fishes were placed inside the smoking kiln with the help of removable wire mesh tray.

Smoking

The lower chamber of the smoking kiln had the facilities of burning saw dust or wood chips on an iron bowl so to produce a continuous and a homogenous hot smoke. The temperature inside the smoke chamber was recorded at every 10 minutes and the temperature was controlled by the exit cover of the kiln. During the smoking procedure, the smoke temperature inside the smoking kiln was recorded by a sensitive thermometer. The desired temperature ranged between 50-55^{0}C and was maintained manually by controlling the outlet of the smoking chamber. During the smoking operation fishes were turned upside down in the middle period, by using a corrosion free metal element to make the sample smooth and steady in texture and appearance. After smoking, products were cooled for 15-20 minutes at room temperature. Cooling firmed the muscle and facilitated to prevent breaking of smoked fish.

Drying Procedure

The smoked fish was than dried with the help of a Ring tunnel dryer. The structure of Ring tunnel is very simple compared to other dryer (Solar tunnel dryer, Rotary dryer). A 6.5 to 7 feet long piece of bamboo was splitted equally in to 6-7 parts the rear end unsplitted. Several rings (outer rings) 1-15 ft diameters were made of split bamboo were tied up with individual splitted bamboo. Round thin-meshed sieves (*chaluni*) made of bamboo inside the shreds were set and tied at regular distance to give a shape of a robust torpedo (Plate 1). The vertical distance between two sieves was about 1.7 feet. Four sieves with similar manner were fixed inside the circular frame. Small fishes were spreaded on these sieves inside the ring tunnel carefully for sun drying. The tunnel was installed vertically keeping the unspiltted bamboo at the upside. After loading of the trays with smoked fish all sides of the tunnel was carefully covered by the polythene except a little whole was kept open at the upper portion of the ring tunnel for the exit of warm air.

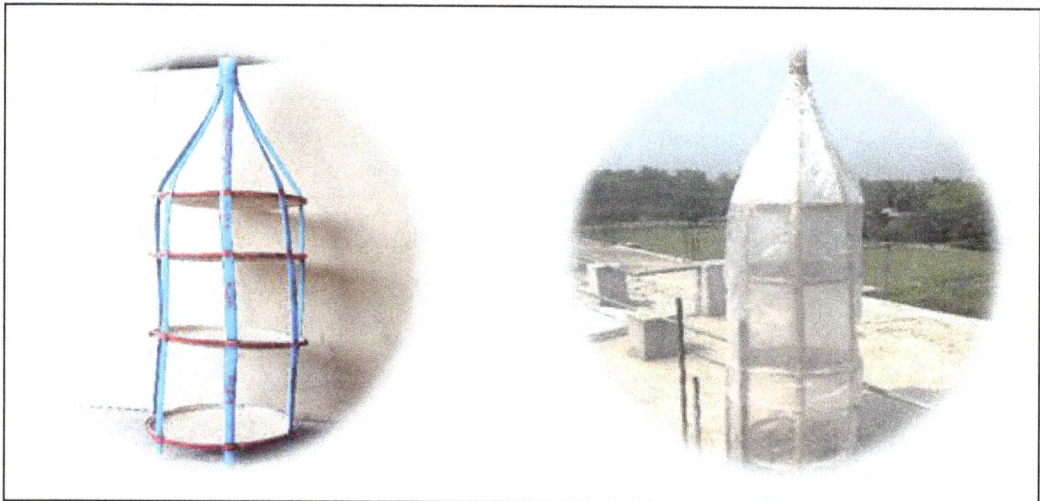

Plate 1. Ring tunnel dryer (unloaded)   **Plate 2.** Ring tunnel dryer (operational view)

The ring tunnel containing the smoked fish was placed at the eastern side of the roof of Prof. Aminul Haque building in the faculty of Fisheries, Bangladesh Agricultural University, Mymensingh to ensure day-long continuous sunshine on the fish in ring tunnels (Plate 2). Temperature inside and outside the ring tunnels during the whole day of drying period of batashi fishes were recorded carefully with the help of a sensitive thermometer. The process continued till the completion of drying. For complete drying of batashi took about 3days (18-20) hours until the final products had moisture level of less than around 20%.

Packaging, Leveling and Storage of the salt-smoke-dried fish products for quality and shelf life study

The smoke-dried fishes were packed in polythene bag and was sealed using an electrical sealing machine (PFS-300) to prevent moisture absorption. Each bag contained 100g salt-smoke-dried fishes with proper labeling to be used for further analysis. The packaged fishes were divided into two parts for storage-one part was kept at refrigeration temperature (4^{0}C) and other portion kept in ambient temperature (26-28^{0}C).

Plate 3. Salt-smoked batashi fish product   **Plate 4.** Salt-smoke-dried batashi fish product

Analysis

For quality and shelf life analysis sampling was done on every 15 days interval for the sample kept at ambient and refrigeration temperature. The nutritional quality on proximate composition (moisture, crude protein, lipid and ash) of the fresh and salt-smoked, salt-smoked dried and control dried fish was carried out in triplicate according to AOAC 1990. Total bacterial count of fresh, salt-smoked, salt-smoke-dried and control dried fish samples were done on nutrient agar media by Standard Plate Count (SPC) method. This method consists of growing the bacteria in a nutrient culture petridish and counting the colonies in the petridish which developed at defined incubation period (35-37°C for 48 hours).

RESULTS AND DISCUSSION

In the present study, it shows the concentration of biochemical composition and microbial populations in fresh, smoked, salt-smoke-dried and control dried batashi fish samples.

Result of Nutritional Quality Analysis

The values of crude protein, lipid and ash have been expressed as dry weight basis. There was a gradual increase in the moisture content of these two types of products with increasing storage period in both ambient and refrigeration condition (Table 1 and 2). This result coincides with the findings of Nahid *et al.* (2014, 2016) and Mazumder *et al.* (2008) in *A. mola, P. chola, G. chapra* and in *P. atherinoides*. On the other hand moisture content increased slowly for the products stored at refrigeration temperature and the values increased from 18.10 (1st day) to 21.42% (60 day) for salt-smoke-dried batashi and from 18.54 (1st day) to 22.86% (60 day) (Table-2) Such increase in moisture content these types of products might be due to the absorption of moisture from surrounding since there was no re-drying during storage period (Darmola *et al.* 2007). The moisture content of smoke-dried chapila, kakila and baim fish rose 12.36 to 14.05%, 11.69 to 13.93% and 8.22 to 15.37% respectively (Nahid *et al.*, 2016) during storage at room temperature which is more or less similar with the present study. The shelf life of both the two types of showed more or less similar with the observation of Jallow who stated that fish at 10-15% moisture content reportedly had a shelf life of 3-9 months when stored properly (Jallow, 1995).

The crude protein content of salt-smoke-dried (SSD) and control dried (CD) product on 0 day was 65.93 and 65.50 %(dry matter basis) respectively. This concentration was resulted from the loss of moisture by the smoking process as opined by Koral *et al.* (2009). Protein decomposes with passing time (Ghezala, 1994).The protein content of salt-smoke-dried (SSD) and control dried (CD) batashi decreased to 65.0 and 58.96%, respectively on dry matter basis after two month storage at ambient temperature. After two month stored at refrigeration temperature the protein content of salt-smoke-dried control dried batashi showed a small decrease to 65.03 and 62.80% on dry matter basis, respectively. In storage condition, the protein content decreased significantly with the time due to water soluble protein diffused out to the surrounding for exosmosis (Hasan *et al.*, 2013). Darmola *et al.* (2013) found the decreasing trend of protein content in hot smoked *Clarius gariepinus* during storage period which is more or less similar with the present findings.

Table 1. Proximate composition (dry weight basis), TVB-N and bacterial count of Fresh, Salt-Smoked, Salt-Smoke-Dried and Control Dried Batashi Fish Stored at Ambient Temperature (26-28) ^{0}C

Day of Observation	Products Name	Moisture (%)	Protein (%)	Lipid (%)	Ash (%)	TVB-N (mg/100g)	SPC (CFU/g.)
0	Fresh fish	75.81±0.18	64.65±0.70	15.71±0.29	13.23±0.57	1.80±0.32	2.72×10^5
0	Salt-smoked fish	61.65±0.13	65.95±0.29	16.84±0.29	12.57±0.16	4.25±0.26	2.64×10^4
0	SSD	18.10±0.11	65.93±0.17	17.09±0.17	16.90±0.06	6.14±0.18	1.14×10^4
0	CD	18.54±0.18	65.50±0.21	18.02±0.17	16.27±0.50	7.05±0.46	1.88×10^4
15	SSD	18.86±0.17	65.54±0.07	16.91±0.25	16.69±0.16	8.82±0.38	1.8×10^4
15	CD	18.78±0.30	65.08±0.14	17.88±0.44	15.63±0.34	9.46±0.46	2.64×10^4
30	SSD	20.14±0.06	65.44±0.04	16.43±0.08	16.23±0.04	11.68±0.42	2.32×10^4
30	CD	20.16±0.22	64.98±0.18	17.46±0.55	15.21±0.11	13.46±0.14	3.9×10^4
45	SSD	21.64±0.03	65.08±0.40	16.18±0.26	16.08±0.09	15.76±0.54	2.9×10^4
45	CD	21.10±0.09	64.36±0.42	17.11±0.58	15.03±0.30	16.14±0.33	4.68×10^4
60	SSD	23.26±0.06	65.0±0.09	15.9±0.33	15.95±0.22	17.94±0.21	4.2×10^4
60	CD	26.30±0.34	58.96±0.84	16.01±0.08	14.49±0.39	19.42±0.42	2.5×10^5

Table 2. Proximate composition (dry weight basis), TVB-N and bacterial count of fresh, salt-smoked, salt-smoke-dried and control dried batashi fish stored at refrigeration temperature (4) ^{0}c

Day of Observation	Products Name	Moisture (%)	Protein (%)	Lipid (%)	Ash (%)	TVB-N (mg/100g)	SPC (CFU/g.)
0	Fresh fish	75.81±0.18	64.65±0.70	15.71±0.29	13.23±0.57	1.80±0.32	2.72×10^5
0	Salt-smoked fish	61.65±0.13	65.95±0.29	16.84±0.29	12.57±0.16	4.25±0.26	2.64×10^4
0	SSD	18.10±0.11	65.93±0.17	16.84±0.29	12.57±0.16	5.86±0.22	1.14×10^4
0	CD	18.54±0.18	65.50±0.21	17.09±0.17	16.90±0.06	6.88±0.05	1.88×10^4
15	SSD	18.32±0.14	65.82±0.13	18.02±0.17	16.27±0.50	6.24±0.11	1.36×10^4
15	CD	19.87±0.14	64.89±0.52	16.99±0.28	16.75±0.09	7.24±0.18	1.96×10^4
30	SSD	19.20±0.21	65.64±0.20	17.82±0.26	16.07±0.44	7.16±0.23	1.64×10^4
30	CD	20.28±0.22	64.15±0.40	16.34±0.17	16.21±0.24	9.34±0.26	2.84×10^4
45	SSD	20.05±0.17	65.32±0.08	17.53±0.30	15.08±0.26	8.96±0.14	2.2×10^4
45	CD	21.86±0.48	63.75±0.49	16.24±0.35	16.16±0.39	11.45±0.14	3.10×10^4
60	SSD	21.42±0.25	65.03±0.17	17.28±0.28	14.89±0.59	11.81±0.39	2.42×10^4
60	CD	22.86±0.58	62.80±0.65	16.16±0.45	16.12±0.09	15.08±0.24	1.16×10^5

##SSD=Salt-smoke-dried, CD=Control dried product (Values are shown on dry matter basis)

The initial lipid content of salt-smoke-dried batashi was high compared to the fresh fish. After smoke-drying, there was an increase in fat content and this variation could be result of evaporation of moisture contents which is in agreement with the previous research works of Ogbonnaya and Shaba (2009), Darmola et al. (2007), Bouriga et al. (2012) and Biligin et al. (2008). Lipid content of salt-smoke-dried (SSD) and control dried (CD) batashi was 17.09 and 18.02%, respectively on dry matter basis, immediately after preparation. During the study period (60 days) lipid content of the samples slowly decreased from their initial values, this was due to the inverse relationship between moisture and fat content. After two month storage at ambient temperature lipid content of salt-smoke-dried and control dried batashi was decreased to 15.9 and 16.01%, respectively on dry matter basis. On the other hand, products stored at refrigeration temperature lipid content of salt-smoke-dried and control dried batashi was found to be decreased to 16.16 and 16.77%, respectively on dry matter basis. Similar decreasing trend of lipid content was found in salt and garlic treated smoke-dried chapila and guchi baim (Nahid et al., 2014). Usually moisture and fat contents in fish flesh are inversely related and there sum is approximately 80% (FAO, 1999) which is well defined in this research findings.

Ash content of salt-smoke-dried and control dried fish samples was higher than that of fresh fish. Salan et al., (2006) observed increase of ash content in smoked *C. gariepinus* due to the loss of humidity and the significant reduction of moisture content during smoking and drying. At 1st day of observation ash of salt-smoke-dried and control dried batashi was 16.90 and 16.27%, respectively. After two month storage at ambient temperature ash content of salt-smoke-dried and control dried batashi was decreased to 15.95 and 14.49, respectively. On the other hand, samples kept 60 days at refrigeration temperature, ash content of salt-smoke-dried (SSD) and control dried (CD) batashi was decreased to 16.12 and 14.35%, (on dry matter basis), respectively. Smaller sized fish species has higher ash content due to the higher bone to flesh ratio (Darmola et al., 2007).

TVB-N (mg/100g) value

In this study the higher value of TVB-N were reported in fresh salt-smoke-dried, control dried batashi compared with fresh fish. There was continuous increase in TVB-N value of all smoke-dried products throughout the storage period (Figure 1) which could be due to gradual degradation of the initial protein to more volatile product such as total base nitrogen (Darmola *et al.,* 2007). The TVB-N value for the products stored at ambient temperature was varied between 6.14 (1st day) to 17.94mg/100g (60 day) for salt-smoke-dried batashi, 7.05 (1st day) to 19.42mg/100g (60 day) for control dried batashi. On the other hand products stored at refrigeration temperature the TVB-N values were found varied between 6.14 (1st day) to 11.20 mg/100g (60 day) for salt-smoke-dried batashi, 7.05 (1st day) to 12.84 mg/100g (60 day) for control dried batashi. Increase in final values of TVB-N in this study was similar to the result of Hasan *et al.* (2006) who reported that the TVB-N values of the dried products from rotary dryer ranged from 10.64 mg/100g to 17.52 mg/100g with lowest in mola dried in rotary dryer in room condition and highest in tengra dried in rotary dryer under direct sunlight.

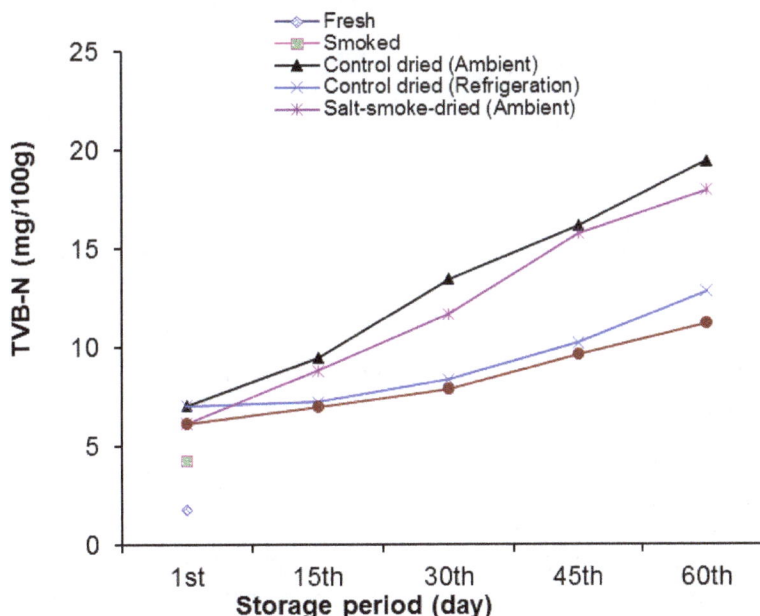

Figure 1. Changing pattern of TVB-N content (mg/100g) of batashi with different treatments stored at ambient (26-28°C) and refrigeration (4°C) temperature

Pearson (1982) and Connell (1995) recommended that the limit of acceptability of fish is 20 to 30 mgN/100 g, while Kirk and Sawyer (1991) suggested a value of 30 to 40 mgN/100 g as the upper limit. Increase in final values of TVB-N in this present research work is similar with the other researcher (Abolagba, 2008 and Trinidad, 1986).

Microbiological Quality Analysis

Bacterial load in fresh batashi was found 2.72×10^5 CFU/g, after smoking bacterial load reduced to 2.64×10^4 CFU/g. This is due to the bacteriostatic and bactericidal effect of wood smoke, heat generation from smoke and also reduction of moisture content in fish body. The initial bacterial load was 1.14×10^4 and 1.88×10^4 CFU/g for salt-smoke-dried and control dried batashi, respectively. The bacterial load increased slowly with the progress of storage time and the value of Standard Plate Count (SPC) at the 60 day of storage at ambient temperature was 4.2×10^4 and 2.5×10^5CFU/g for salt-smoke-dried and control dried batashi, respectively (Figure 2). On the other hand value of SPC for the products stored at refrigeration temperature were increased to $2.42\times\times10^4$ and 1.16×10^5 CFU/g for salt-smoke-dried and control dried batashi, respectively. The CFU/g of smoke-dried fish samples increased with increase in storage period due to growth and multiplication of the microbes (Bilgin et al., 2008). Zaki et al. (1976) that reported the total bacterial count decreased after drying, owing to the high salt content and the lack of enough free water in fish tissues.

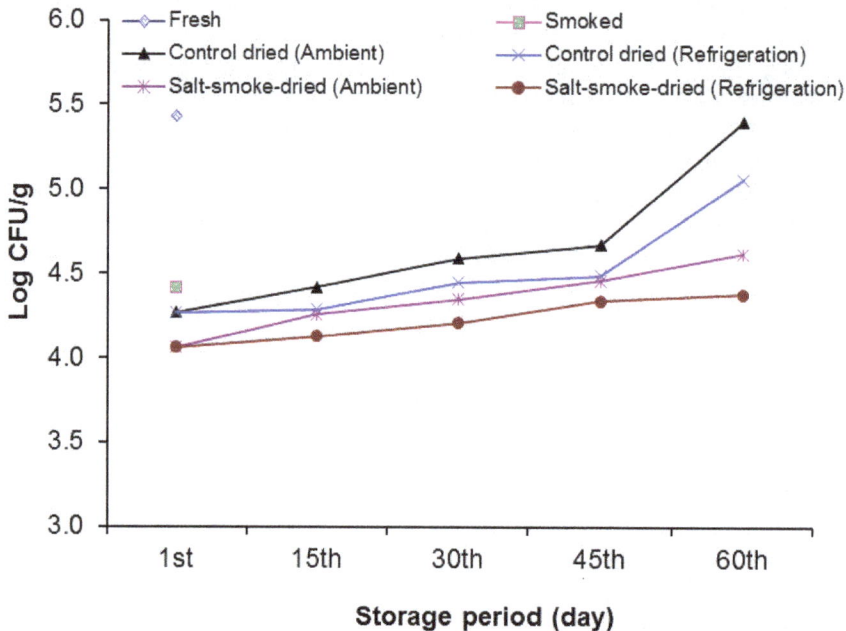

Figure 2. Changing pattern of bacterial load (log CFU/g) of batashi with different treatments stored at ambient (26-28°C) and refrigeration (4 °C) temperature

Similar findings has also been observed by Kamruzzaman (1992) where bacterial count of commercially dried freshwater fish samples ranged from 1.84×10^4 to 5.3×10^6 CFU/g. Hasan et al. (2006) showed that much higher bacterial load of traditional, rotary and solar tunnel dried products (mola, tengra and katcki), were in the range of 1.43×10^8 to 2.89×10^8 CFU/g, 1.91×10^8 to 2.84×10^8 CFU/g and 1.95×10^8 to 2.59×10^8 CFU/g, respectively. This experiment with batashi also provides more or less similar result with the findings of the above studies.

CONCLUSION

Salt-smoke-dried SIS (batashi) produced in smoking kiln and ring tunnel dryer can be stored in polythene package at ambient and refrigeration temperature for more than two months without any quality loss, however the salt-smoke-dried SIS products stored in refrigeration temperature can provide much longer shelf-life by minimizing the moisture uptake and bacterial load. It was observed that the use of salt and smoke comparatively had a special smoky flavor with good texture in the product. The product made by this process showed a better hygienic aspect by shortening the drying period of fish. The saving of time and improvement of the quality in the salt-smoke-dried fish process will help the poor fisher folks getting better price for their products but also enhance consumer preference in the local market as well as export market. One of the major findings from the present study was the effect of salt and smoke on the insects (flies). Fish treated with only salt was seen to be infested by few insects, whereas no infestation was seen in fish treated with salt and smoke on the other hand fish which were not treated with salt and smoke (control), the product was infested by insect during drying in open place.

COMPETING INTEREST

The authors declare no competing interests.

ACKNOWLEDGEMENT

The authors would like to thank BAURES, Bangladesh Agricultural University, Mymensingh-2202, Bangladesh for funding to carry out this research work successfully.

REFERENCES

1. Abolagba OJ and SJ Osifo, 2008. The effect of smoking on the chemical composition and keeping qualities of cat fish *Heterbranchus bidorsalis* using two energy sources. Journal of Agriculture, Forestry and Fisheries, 5: 27-30
2. AOAC, 1990. Official method of analysis. Association of Official Agricultural Chemists W. Horwitz (Editor) 12~ ED. Washington.
3. Ashano MO and OE Ajayi, 2003. Effect of processing and storage methods on the shelf life and incidence of insect pest on smoked fish. Global Journal of Pure and Applied Science, 9: 13-15.
4. Bilgin F, M Unliisayin, L Izci and A Gunlu, 2008. The determination of the shelf life and some nutritional components of gilthead seabream (*Sparus aurata* L., 1758) after Cold and Hot Smoking. Turkish Journal of Veterinary and Animal Sciences, 32: 49-56.
5. Bouriga N, IH Ben, M Gammoudi, E Faureand M Trabelsi, 2012. Effect of smoking method on biochemical and microbiological quality of Nile Tilapia (*Oreochromis niloticus*). American journal of food echnology, 7: 679-689.
6. Clucas IJ, 1981. Fish handling, preservation and processing in the tropics. 33-36.
7. Connell JJ, 1995. Control of fish quality. 4th Edition. Fishing News Books, Farnham. England.
8. Daramola JA, CT Kester and OO Allo, 2013. Biochemical evaluation of hot-smoked African catfish (*C. gariepinus*) sampled from Sango and Ota market in Ogun State. 382 pp.
9. Daramola JA, EA Fasakin and EO Adeparusi, 2007. Changes in physicochemical and sensory characteristics of smoke-dried fish species stored at ambient temperature. African Journal of Food, Agriculture, Nutrition and Development, 7: 169-183.
10. FAO, 1999. World production of fish, crustaceans and mollusks by major fishing areas. Fisheries Information Data and Statistics unit (FIDI), Fisheries Department, FAO Rome.pp.33.
11. Ghezala S, 1994. New packaging technology for seafood preservation shelf life extension and pathogen control". in: fisheries processing biotechnological applications. A.M. Martin (ed.). Chapman Hall: London, UK. 83-110.

12. Hasan MM, FH Shikha, MI Hossain, M Kamal, MN Islam and MA Wahab, 2006. Quality assessments of traditional, rotary and solar tunnel dried small indigenous fish products. Bangladesh. Journal of Fisheries Research, 10: 73-84.

13. Ja11ow AM, 1995. Contribution of improved Chorkor Oven to Artisanal fish smoking in the Gambia. Workshop on seeking improvement in fish technology in West Africa IDAF Technical Report. No. 66.

14. Kamruzzaman AKM, 1992. Qualitative evaluation of some commercial dried fish products of Bangladesh. MS thesis, submitted to Bangladesh Agricultural University, Mymensingh, Bangladesh, 37.

15. Kirk RS and R Sawyer, 1991.Nitrogen determination. Pearson's composition and analysis of foods. Longman Scientific Publisher: London, UK. 29-36.

16. Koral S, S Kose and B Tufan, 2009. Investigating the quality changes of raw and hot smoked garfish (*Belone belone euxini,* Gunther, 1866) at ambient and refrigerated temperatures. Turkish Journal of Fisheries Aquatic and Sciences, 9: 53-58.

17. Kumolu-Johnson CA, NF Aladetohun and PE Ndimele, 2009. The effects of smoking on the nutritional qualities and shelf-life of *Clarias gariepinus* (BURCHELL 1822). African Journal of Biotechnology, 9: 73-76.

18. Mansur MA, MA Sayed, SC Chakraborty, MN Uddin and MNA Khan, 1998. Influence of spawning on the sensory, biochemical and bacteriological characters of dry salted Hilsa (*Hilsa ilisha*) fish at ambient temperature. Bangladesh Journal Fisheries, 21: 65-72.

19. Mazumder MSA, MM Rahman, ATA Ahmed, M Begum and MA Hossain, 2008: proximate composition of some small indigenous fish species (SIS) in *Bangladesh.* International Journal of Sustainable Crop Production, 3: 18-23.

20. Nahid MN, GA Latifa, FB Farid, M Begum, 2014. Evaluation of biochemical composition of salt and garlic treated smoke-dried chapila (*Gudusia chapra* Hamilton, 1822) and Kaika (*Xenentodon cancila,* Hamilton-Buchanan, 1822) fish at laboratory condition (27-31^{0}C). Research Journal of Animal, Veterinary and Fishery Sciences, 2: 10-15.

21. Nahid MN, GA Latifa, SC Chakraborty, FB Farid and M Begum, 2016. Shelf-life quality of smoke-dried freshwater SIS fish; chapila (*Gudusia chapra*, Hamilton-Buchanan) kakila (*Xenentodon cancila*, Hamilton-Buchanan; 1822) stored at laboratory condition (26-31^{0}C). Journal of Agriculture and Veterinary Science, 9: 23-32.

22. Ogbonnaya C and IM Shaba, 2009. Effects of drying methods on proximate composition of cat fish (*Clarius gadepinus*). World Journal of Agricultural Sciences, 1: 114-116.

23. Pearson D, 1982. The chemical analysis of foods. Churchill Livingstone, Edinburgh, London and New York.

24. Salan OE, AG Julian and O Marilia, 2006. Use of smoking to add value to Salmoned trout. Brazillian Archives of Biology and Technology, 49: 57-62.

25. Salim GM, MN Islam and SC Chakraborty, 2007. Shelf-life and bacteriology of smoked fish prepared from Thai pangus (*Pangusius hypophthahnus*). Progressive Agriculture, 18: 199-207.

26. Trinidad LM and MH Estrada, 1986. `Effect of raw material freshness on the quality of smoked tilapia *(Oreochromis nfloticus)"*. In: J.L. Maclean, L.B. Dixon, and L.V. Hosilus (eds.). The First Asian Fisheries Forum. Manilla, Philippines. 471-472.

27. Zaki MS, YM Hassan and EHA Rahma, 1976. Technological studies on the dehydration of the nile bolti fish (*Tilapia nilotica*) Nahrung, 20: 467-474.

14

ECONOMICS OF MIXED RICE-FISH FARMING IN SOUTH-WEST REGION OF BANGLADESH

Apurba Roy

Department of Economics, Faculty of Social Sciences, University of Barisal, Barisal-8200, Bangladesh

*Corresponding author: Apurba Roy; E-mail: apurbo.roy2@gmail.com

RTICLE INFO	ABSTRACT
Key words Mixed farming, Production efficiency, Scale to sale	The present research aims at investigating the economic performance of mixed rice-fish farming in south-west region of Bangladesh. In order to carry out the research objective descriptive statistics, profit function, Cobb-Douglas form of multiple linear regression model and t-test approaches have been applied. The study area has been selected using multi-stage sampling technique and convenient sampling method has been utilized to select the sample. In-depth interview technique has been employed to collect primary data by using pretested questionnaire from the samples. Results from descriptive statistics show that the average annual return on mixed rice-fish farming is BDT 56326.45 more than mono rice farming as well as production efficiency of mixed rice-fish farming is also found higher than mono rice farming. Besides, mixed rice-fish farming experiences increasing return to scale, whereas, mono rice farming undergoes decreasing return to scale. Moreover, test of hypothesis provides statistical evidence that mixed rice-fish farming is more profitable than mono rice farming in the study area.

INTRODUCTION

Bangladesh is a country of agriculture situated in south-east Asia. It is one of the highly populated least developed countries in the world with an area of 144000 square kilometers (Ahmed et al., 2011). Agriculture sector is the driving forces of economic growth of Bangladesh contributing 19.41 percent to the gross domestic product (GDP) in fiscal year 2011-2012 and providing a space of occupation for 47.5 percent of the total labor force in the same year (BBS, 2013). Rice and fish are two important products of agriculture becoming the part and parcel of the people of the country. As the main crop of Bangladesh, rice is grown around 29 million tons per year as annual production, whereas, the annual production of fish is around 2.70 million tons (Ahmed and Garnett, 2011). With a population of 16.4 crore, the demand of both rice and fish is increasing continuously in the country every year (Ahmed et al., 2011). The average size of cultivable land holding is decreasing in Bangladesh forcing the farmers to bear high risk and cost to maximize crop production within a shortest possible time due to high pressure from increasing population. In this backdrop, practice of mixed rice-fish farming can be a way to tackle the growing demand for food by utilizing farm resources with least time and cost of production. The exercise of rice-fish cultivation can also be source of decent supply of carbohydrate and animal protein (Ahmed and Garnett, 2011; Mamun et al., 2012; Roy et al., 2013).

The total amount of rice cultivated area in Bangladesh is near about 10 million hectare which covers about 75 percent of the total cropped areas (Ahmed and Garnett, 2011; Roy et al., 2013). The country's total production of rice is 344.30 lakh metric tons and in case of fish, it remains 32.22 lakh metric tons in fiscal year 2011-2012. The available capacities for producing rice and fish are not being utilized fully although there exists potential to use them. Each year, a lot of food stuffs are imported to meet up the excess demand. In fiscal year 2011-2012, around 22.36 lakh metric tons food stuffs have been imported by both government and non-government sector (BBS, 2013). In this situation, producing rice-fish simultaneously in the same field paves the way for optimum resource utilization for maximum food production. In terms of cost and benefit, mixed rice-fish farming is also being preferred by farmers than mono rice farming in the recent times. One of the reasons may be that integration of fish and rice leads to lower production costs, where sails, insects, pests, and other harmful flies are caught and eaten by the fish.

Justification of the research

Extensive literature on mixed farming is still very few. In addition, studies on economic analysis on mixed rice-fish farming over mono rice production practice is also scarce. A little number of studies have been found on mixed farming, specifically on rice-fish mixed farming in Bangladesh perspective. Among them, a study conducted by Hasanuzzaman and the others investigate practice and economics of freshwater prawn farming in seasonally saline rice field at Shyamnagore Upazilla in Satkhira District of Bangladesh indicating that the culture of prawn in seasonally saline paddy field is economically viable (Hasanuzzaman et al., 2011). Another study explores the fish-paddy crop rotation practice at the farmer's pond at Sadar Upazilla of Bagerhat District, in south-west coastal region of Bangladesh describing that fish-paddy crop rotation system enhances the fertility of the ponds (Roy et al., 2013). However, another study conducted in Mymensingh District of northcentral Bangladesh provides evidence that integrated rice-fish farming can help Bangladesh keep pace with the current demand for food through rice and fish production (Ahmed and Garnett, 2011). On the other hand, Integrated Farming System (IFS) can eradicate high degree of risk and uncertainty because of seasonal, irregular and uncertain income and employment to the farmers by not only solving most of the existing economic and even ecological problems, but also provide other household needs like fuel, fertilizer and feed, along with increasing productivity of the farm manifold (Mamun et al., 2012). The objective of the research is to assess the economic performance of mixed rice-fish farming over mono rice farming at Dighalia Upazila in Khulna District of Bangladesh. Besides, the study also intends to estimate the production function of the two group of farmers to measure the production efficiency and returns to scale.

METHODS AND MATERIALS

Study area

Present study has been conducted in the south-east region of Bangladesh. Multistage sampling technique has used to select the study area. In the first stage, Khulna District has been chosen for purpose of the study from total of 64 districts in Bangladesh; because, this district is situated in the south-west region of Bangladesh. In the second stage, Dighalia Upazila, shown in Figure 1, has been selected from Khulna Districts because most of the farmers in Dighalia Upazila are engaged in rice cultivation and fish cultivation. At the third step, Gazirhat Union has been selected purposively as the farmers in this area are engaged in both rice-fish farming and mono rice production under double cropping system all the year round. At the last stage, a total of 8 villages of the union have been selected randomly.

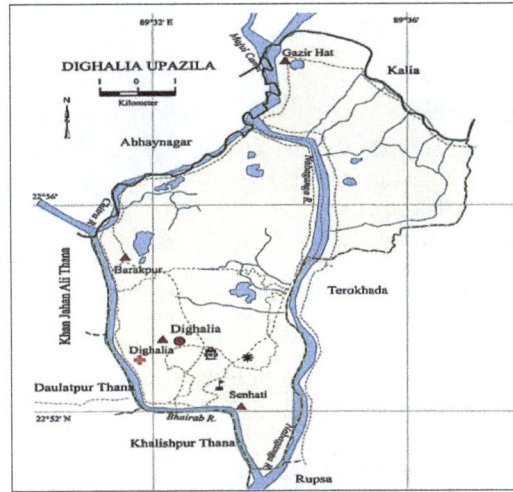

Figure 1. The study area

Source: Asiatic Society of Bangladesh (2006)

Sampling and data collection

For study purpose, farmers who produce only rice and farmers who produce rice-fish have been considered as population. Existing literature survey and a pilot field survey confirm that double cropping system is in operational for rice cultivation in the study area. Generally, cropping season for Boro rice ranges from November to April and duration of Aman rice is June to September (Ahmed and Garnett, 2011; Ahmed et al., 2011; Hasanuzzaman et al., 2011). On the other hand, two widely practiced rice-fish cultivation systems are concurrent culture (integrated) – growing the fish together with the rice in the same area – and rotational culture (alternate) – where the rice and fish are grown at different times (Ahmed et al., 2011; Hasanuzzaman et al., 2011; Roy et al., 2013). Time frame for integrated rice-fish farming is July to November and December to April for Boro rice production. On the other hand, alternate rice-fish farming involves producing of fish in the monsoon from June to December and cultivation of Boro rice from November to April.

In case of study area, quick field survey reveals that mono rice farmers produce Boro and Aman in two cropping season, on the other hand, most of the rice-fish farmers are practicing alternate culture due to the lack of enough water and fish prawn in the dry season. So, the samples of the present study include 30 mono rice farmers and 30 alternate rice-fish producing farmers from the time period of June 2015 to April 2016. The sample has been selected purposively for the study.

For accomplishing the study, data have been collected from both primary and secondary sources. In order to collect primary data, well-structured questionnaire, pretested and revised on the basis of findings and experiences originated at the time of pilot survey has been used. Primary cross-section data have been collected through questionnaire survey at the time of final field survey. Besides, necessary secondary data have been collected from published book, journals, working papers, internet etc.

Data analysis and model specification

In order to analysis data, some mathematical tools have been applied in the study. Profit function has been used to identify the net profit of the farmers from mono rice and rice-fish cultivation, and a Cobb-Douglas production function has been applied to estimate the production efficiency of rice-fish farming and mono rice farming in association with measuring the returns to scale from both production function. In the Table 1, description of the variables used has been mentioned with measurement unit.

Table 1. Description of the variables

SL.	Variable name	Symbol	Measurement unit
i.	Total rice production	Y_r	Kg per acre per cropping season
ii.	Rice seed	X_{r1}	Kg per acre per cropping season
iii.	Fertilizer	X_{r2}	Kg per acre per cropping season (rice production)
iv.	Labor	X_{r3}	No. of labor man-day per acre per cropping season (rice production)
v.	Farm size	X_{r4}	In acre (rice production)
vi.	Total fish production	Y_f	Kg per acre per cropping season
vii.	Fish stocking	X_{f1}	Kg per acre per cropping season
viii.	Fish feed	X_{f2}	Kg per acre per cropping season
ix.	Fertilizer	X_{f3}	Kg per acre per cropping season (fish production)
x.	Labor	X_{f4}	No. of labor man-day per acre per cropping season (fish production)
xi.	Farm size	X_{f5}	In acre (fish production)

Source: Author's compilation, 2016

Estimation of profit function

For calculating profit level of the farmers, the following general profit functions have been used.

$$\Pi = TR - TC \text{---(1)}$$

Where,

Π = Net profit

TR = Total revenue earned or, [TR = P*Q, where, P = price level and Q = quantity produced]

TC = Total cost incurred or, [TC = TFC + TVC, where, TFC = Total fixed cost and TVC = Total variable cost]

i. Profit function for mono rice production and mixed rice-fish production measured in (BDT/acre/year)

$$\Pi_{ij} = TR_{ij} - TC_{ij} \text{--(2)}$$

Where,

Π = Net profit, TR = Total revenue, TC = Total cost, i = Profit function for mono rice production, j = Profit function for mixed rice-fish production

Estimation of production efficiency model

Production efficiency of rice-fish production and mono rice production process has been estimated to observe the efficiency level of the farmers. Econometric model of Cobb-Douglas production function is widely used to estimate production efficiency in agriculture. Since, alternative mixed rice-fish technique is prevalent in the study area, separate equation has been applied to estimate the efficiency level of alternate mixed rice-fish and mono rice production.

i. Production efficiency of rice production under alternate mixed farming

$$\ln Y_{ri} = \beta_0 + \beta_1 \ln X_{r1i} + \beta_2 \ln X_{r2i} + \beta_3 \ln X_{r3i} + \beta_4 \ln X_{r4i} + \ln U_i \text{--------------------------------(3)}$$

Where, Y_r = Total rice production, X_{r1} = Rice seed, X_{r2} = Fertilizer, X_{r3} = Labor, X_{r4} = Farm size, U_i = Error term, β_0 = Constant parameter in the equation, mathematically interpreted as the intercept, β's = Coefficient of the relevant variables, i = observation on i^{th} farmer (1, 2, 3 .... 30)

ii. Production efficiency of fish production under alternate mixed farming

$$\ln Y_{fi} = \beta_0 + \beta_1 \ln X_{f1i} + \beta_2 \ln X_{f2i} + \beta_3 \ln X_{f3i} + \beta_4 \ln X_{f4i} + \beta_5 \ln X_{f5i} + \ln U_i \text{--------------------(4)}$$

Where, Y_f = Total fish production, X_{f1} = Fish stocking, X_{f2} = Fish feed, X_{f3} = Fertilizer, X_{f4} = Labor, X_{f5} = Farm size, U_i = Error term, β_0 = Constant parameter in the equation, mathematically interpreted as the intercept, β's = Coefficient of the relevant variables, i = observation on i^{th} farmer (1, 2, 3 .... 30)

iii. Production efficiency of mono rice production

For estimating mono rice production (Aman season and Boro season), two modified Cobb-Douglas production function has been used as presented in equation (5) and (6).

Production efficiency of Aman rice:

$$\ln Y_{rai} = \beta_0 + \beta_1 \ln X_{r1i} + \beta_2 \ln X_{r2i} + \beta_3 \ln X_{r3i} + \beta_4 \ln X_{r4i} + \ln U_i \text{--------------------------------(5)}$$

Production efficiency of Boro rice:

$$\ln Y_{rbi} = \beta_0 + \beta_1 \ln X_{r1i} + \beta_2 \ln X_{r2i} + \beta_3 \ln X_{r3i} + \beta_4 \ln X_{r4i} + \ln U_i \text{------------------------(6)}$$

Where, Y_{ra} = Total Aman rice production, Y_{rb} = Total Boro rice production, X_{r1} = Rice seed, X_{r2} = Fertilizer, X_{r3} = Labor, X_{r4} = Farm size, U_i = Error term, β_0 = Constant parameter in the equation, mathematically interpreted as the intercept, β's = Coefficient of the relevant variables, i = observation on i^{th} farmer (1, 2, 3 .... 30),

iv. Measurement of returns to scale

Returns to scale have been estimated both for mono rice production and mixed rice-fish production by using following equation adopted from (Ahmed and Garnett, 2011).

$$r = \sum \beta_i \text{--(7)}$$

Where, r = Returns to scale

$\sum \beta_i$ = Sum of all the production coefficients

r < 1 = Decreasing returns to scale

r = 1 = Constant returns to scale

r > 1 = Increasing returns to scale

Hypothesis testing

This study intends to find out the economic difference between mixed rice-fish and mono rice farming in the study area. For this reason, a research hypothesis has been articulated as given below:

Null Hypothesis:

H_0 = There is no difference between mixed rice-fish and mono rice farming in terms of economic returns.

Alternative Hypothesis:

H_1 = There is economic difference between mixed rice-fish and mono rice farming in terms of economic returns.

For data analysis, a number of computer software have been used. At first, collected primary data have been organized and stored by MS Excel 2016. In the second step, mathematical and statistical calculations have been compiled through using STATA 12 and IBM SPSS Statistics 20.

RESULTS AND DISCUSSION

Social profile of the respondents

Social profile of the respondents helps to get general idea of the samples in the study area. Descriptive analysis presented in the Table 2 shows that among the samples average ages of the mixed faming farmers and mono rice farmers are 40 and 42 respectively. Among the mixed farmers, mean education level is 7 years and it is 5 years for the mono rice produces. In addition, there is little difference between the two type of respondents in terms of household size. However, the average farming experience of the former group of farmers is 16 years, whereas, the latter group of farmers have on an average 20 years of farming experience.

Household farm production related information

Farm related information of the household provides significant insight of the production process of the respondents in the study area. To make it clear, cropping season-wise summary statistics of the respondents has been presented in the Table 3. It has been seen from the Table 3 that, Boro production under alternate mixed rice-fish farming system has an average rice production of 4184 kg per acre. In terms of using inputs per acre, on an average, 57 labor man day, 10 kg of rice seed and 308 kg of fertilizer are applied per acre of land. On the other hand, fish production under the same farming system yields on an average 1233 kg of white fish per acre of land in the same area. In order to produce that level of output, it takes on an average 15 labor man day, 195 kg of fish feed, 68 kg of fish prawn and 60 kg of fertilizer. There is variation in the output and inputs employed from respective farmer to farmer as indicated by standard deviation.

On the other hand, Aman production under mono rice production represents different results apart from mixed farming in the same area. It has been found that, average production of Aman per acre of land is 1836 kg. It takes on an average, 12 labor man day, 13 kg rice seed and 51 kg of fertilizer to produce that level of output. However, the production volume of Boro rice under mono rice farming shows different picture. Summary statistics presented in the Table 3 articulates that on an average 3733 kg of rice is produced per acre. To come up with this amount of output, it takes on average, 28 labor man day, 8 kg of rice seed, and 111 kg of fertilizer. There is variation in the output and input level from producer to producer which is shown by standard deviation in the Table 3.

Table 2. Summary of social profile of the respondents

	Variable	Obs.	Mean	Std. Dev.	Min	Max
Alternate mixed rice-fish farmers	Age	30	40.93	10.05	23	61
	Education	30	7.97	2.41	5	12
	Household size	30	5.67	1.84	3	9
	Farming experience	30	16.73	10.66	3	45
Mono rice farmers	Age	30	42.43	10.77	25	65
	Education	30	5	4.28	0	12
	Household size	30	6.03	1.69	3	10
	Farming experience	30	20.87	10.54	5	42

[Note: Obs. = Observation, Std. Dev. = Standard Deviation, Min = Minimum, Max = Maximum]
Source: Author's compilation based on field survey, 2016

Table 3. Summary of farm production information of the respondents

	Variable	Obs.	Mean	Std. Dev.	Min	Max
Boro production [alternate mixed rice-fish farming]	Rice production	30	4184	1456.53	1260	6720
	Farm size	30	1.20	0.376	0.50	1.98
	Labor	30	57.88	25.73	18	150
	Rice seed	30	10.88	3.84	3	18
	Fertilizer	30	308.9	138.30	37	600
Fish production [alternate mixed rice-fish farming]	Fish production	30	1233	471.24	200	2500
	Farm size	30	1.20	0.37	0.50	1.98
	Labor	30	15.85	14.83	3	85
	Fish feed	30	195.23	69.06	22	300
	Fish stocking	30	68.83	39.71	20	150
	Fertilizer	30	60.17	28.05	10	120
Aman production [mono rice farming]	Rice production	30	1836.33	917.11	800	4200
	Farm size	30	1.14	0.41	0.33	1.65
	Labor	30	12.13	3.60	2	20
	Rice seed	30	13.03	3.38	5	18
	Fertilizer	30	51.2	11.90	40	80
Boro production [mono rice farming]	Rice production	30	3733.67	1456.23	950	5600
	Farm size	30	1.14	0.41	0.33	1.65
	Labor	30	28.1	12.19	12	58
	Rice seed	30	8.93	4.03	1	16
	Fertilizer	30	111.73	37.15	40	188

[Note: Obs. = Observation, Std. Dev. = Standard Deviation, Min = Minimum, Max = Maximum]
Source: Author's compilation based on field survey, 2016

Table 4. Profit level of mixed rice-fish and mono rice production

Type of farming	Cost per acre		Gross return per acre		Net return
	Input	BDT	Output	BDT	BDT
Boro production [alternate mixed rice-fish farming]	Labor	10883.33	Rice production	83680.00	
	Rice seed	2144.83	Rice straw	7746.67	
	Fertilizer	6228.67			
	Land preparation	981.33			
	Total	20238.16		91426.67	71188.51
Fish production [alternate mixed rice-fish farming]	Labor	1832.50	Fish production	147960.00	
	Fish stocking	10845.00			
	Fish feed	8013.67			
	Fertilizer	1547.17			
	Lime cost	766.5			
	Total	23004.84		147960.00	124955.2
Aman production [mono rice farming]	Labor	2263.33	Rice production	27545	
	Rice seed	394.07	Rice straw	2366.67	
	Fertilizer	1536			
	Land preparation	1032.5			
	Total	5225.9		29911.67	24685.77
Boro production [mono rice farming]	Labor	5780	Rice production	74673.33	
	Rice seed	1536.00	Rice straw	9802.9	
	Fertilizer	2234.67			
	Land preparation	800.00			
	Irrigation	2237.08			
	Total	12587.75		84476.23	71888.48
Alternated mixed rice-fish farming [rice + fish]/acre/year	43243		196143.71		152900.7
Mono rice farming [Aman + Boro]/acre/year	17813.65		114387.9		96574.25

Note: Inputs (average in kg/acre/cropping season) and costs (average in BDT/acre/cropping season)
Source: Author's compilation based on field survey, 2016

Estimation of profit level of mixed rice-fish and mono rice production

Integrated rice-fish farming provides benefits such as economic, optimum and double utilization of paddy field, where it is possible to produce fish and rice simultaneously without supplying any excess fertilizer, food and water and rice yield also will increase (Noorhosseini-Niyaki and Bagherzadeh-Lakani, 2011). One of the prime motives of any production activity is to earn revenue. In this case, attempts have been taken to figure out the level of revenue earned by the farmers who practice rice-fish mixed farming and mono rice farming in the study area. Net return from both farming type has been calculated separately based on cropping season and combinedly as presented in the Table 4. It has been seen from the Table 4 that, farmers have to spend money on labor, seed, land preparation, fertilizer, fish stock, fish fee etc. On the other hand, they get return from rice production, rice straw etc. Estimated results shown in the Table 4 indicates that on an average, Boro rice production under alternate mixed rice-fish farming incurs a total of BDT 20238 as input cost. On the other hand, it gets return amounted to BDT 91426. In the end, the amount of net return reaches to near about BDT 71188 to the producer. In the same manner, on an average, a total of BDT 124955 is left as net return in fish farming under alternate mixed rice-fish farming in the same area. In that case, the total cost is BDT 23004 and gross return is equivalent to BDT 147960. However, the net return from mono rice farming is a bit different from the previous one. It is clear from the Table 4 that, on an average Aman production under mono rice farming gives net return of BDT 24685 and Boro production renders a sum of BDT 71888 as net return.

To sum up, it is found that on an average, mixed rice-fish producer get on an average BDT 152900 as annual net return. On the other hand, annual average net return is BDT 96574 for the farmers who practice mono rice farming system in the study area. The analyses presented in the Table 4 crystalize that mixed farming is more profitable than mono rice farming the study area. This result is consistent with the findings of (Dey et al., 2013) showing that alternating rice–fish systems provides substantial potential for increasing productivity and farm incomes in Bangladesh. Besides, Nahar (2010) finds that production level from rice-fish system is significantly higher than other farming system and suitable for poor people considering the yields and economics benefits from it.

Table 5. Results from regression analysis

Explanatory variables	Farming type			
	Boro production [alternate mixed rice-fish farming]	Fish production [alternate mixed rice-fish farming]	Aman production [mono rice farming]	Boro production [mono rice farming]
Rice seed	(-0.102) {0.065} [-1.57]	-	(0.671) {0.361} [1.86***]	(0.001) {0.099} [0.01]
Fertilizer	(0.211) {0.064} [3.30*]	(-0.147) {0.189} [-0.78]	(-0.601) {0.390} [-1.54]	(-0.129) {0.210} [-0.62]
Labor	(0.166) {0.099} [1.68]	(-0.140) {0.151} [-0.93]	(-0.195) {0.210} [-0.93]	(-0.009) {0.169} [-0.05]
Farm size	(0.775) {0.159} [4.87*]	(-0.140) {0.356} [-0.39]	(-0.029) {0.249} [-0.12]	(1.054) {0.151} [6.96*]
Fish stocking	-	(0.262) {0.211} [1.24]	-	-
Fish feed	-	(0.471) {0.207} [2.27**]	-	-
Constant	(6.559) {0.653} [10.03]	(4.472) {1.187} [3.76]	(7.707) {1.301} [5.92]	(8.529) {.617} [13.82]
Summary statistics				
R^2	0.91	0.32	0.21	0.78
Observation	30	30	30	30
Mean VIF	2.95	1.70	1.50	2.10
Return to scale Σb_i	1.05	0.31	-0.154	0.917

[Note: Figure in the first bracket indicates coefficient value, figure in the second bracket indicates standard error and figure in the third bracket refers to t-value, * = significant at 1 percent level, ** = significant at 5 percent level, *** = significant at 10 percent level]
Source: Author's compilation based on field survey, 2016

Table 6. Hypothesis testing

Variable	Obs.	Mean	Std. Err.	Std. Dev.
Gross return from mixed rice-fish farming	30	231640	12721.45	69678.22
Gross return from mono rice farming	30	102218.3	5775.811	31635.42
Difference		129421.7	13971.23	
		t = 9.2634		
		Degrees of freedom = 58		

H_0: mean(diff) = 0
H_a: mean(diff) < 0 H_a: mean(diff)! = 0 H_a: mean(diff) > 0
$Pr(T < t) = 1.0000$ $Pr(|T| > |t|) = 0.0000$ $Pr(T > t) = 0.0000$
[Note: Obs. = Observation, Std. Err. = Standard Error, Std. Dev. = Standard Deviation]

Source: Author's compilation based on field survey, 2016

Production efficiency and reruns to scale of mixed rice-fish and mono rice farming

Production efficiency of the respective farming system has been estimated using econometric regression model of Cobb-Douglas production function. Results from alternative mixed rice-fish farming show that coefficient of multiple determination (R^2) ranges from 0.91 to 0.32. It implies that 32 to 91 percent of total variation in the production of mixed rice-fish farms can be explained by the explanatory variables used in the models. In addition, it omits variables that can explain 9 to 68 percent of the model. The estimated coefficients of farm size and fertilizer are 0.77 and 0.06 respectively. This indicates that 1 percent increase in farm size and fertilizer, keeping other thing constant, would increase the production by 0.77 percent and 0.06 percent respectively and it is statistically significant at 1 percent. In case of fish production under the same farming system, estimated coefficient value of fish feed is 0.47 which is significant at 5 percent level. It states that 1 percent increase in input fish feed results in 0.47 percent increase in the fish production. The other variables are found statistically insignificant. On the other hand, the estimated coefficients of R^2 of Aman and Boro rice production under mono rice farming system are 0.21 and 0.78. It refers that 21 to 78 percent of total variation in the production of mono rice farms can be explained by the explanatory variables used in the models, the other thing remaining constant. In addition, it excludes explanatory variables that could explain 22 to 79 percent of the model. However, estimated coefficient of input rice seed is 0.67 indicating 1 percent increase in rice seed would increase Aman rice production by 0.67 percent and it is statistically significant at 10 percent level. Besides, estimated coefficient of farm size of Boro rice production is 1.05 meaning 1 percent increase in farm size would increase Boro rice production by 1.05 percent, which is statistically significant at 1 percent level. The other variables are found statistically insignificant.

The summation of all the production coefficients (Σb_i) in the alternate rice-fish mixed farming system is 1.36. This result is closer to the findings of (Ahmed and Garnett, 2011), who have revealed that return to scale of alternate rice-fish mixed farming is equivalent to 1.41. Since it is greater than 1, the returns are supposed to increase with the increase in inputs. For instance, if all the inputs mentioned in the model are increased by 1 percent, farm production will increase by 1.36 percent in the alternate rice-fish mixed farming system. This results suggests that there is still scope to increase production under the alternate mixed farming system in the study area by increasing the amount of inputs. On the other hand, the summation of all the production coefficients (Σb_i) in the mono rice farming system is 0.763. Since it is smaller than 1, the returns are supposed to decease with the increase in inputs in the study area.

Hypothesis testing

The result of hypothesis testing has been mentioned in the Table 6. It is seen from Table 6 that, there is statistically significant difference between return of mixed rice-fish farming and mono rice farming in the study area. The mean difference of gross return between the two-farming system is found equivalent to BDT 129421.70. The probability of mean difference to be equal to zero or greater than zero is significant at 1 percent level. So, the null hypothesis implying there is no difference of economic return from mixed rice-fish farming and mono rice farming is rejected at 1 percent level of significance. This estimate provides significant insight and proof that mixed rice-fish farming is more profitable and economically beneficial than mono rice farming in the study area.

CONCLUSION

The study intends to analyze the economic performance of mixed rice-fish farming and mono rice farming in the study area. Results from the descriptive statistics show that average ages of the mixed faming farmers and mono rice farmers are over 40. However, the average farming experience of the two group of farmers' ranges from 16-20 years. Farm related information of the farmer indicates that average Boro rice and fish production under alternate mixed rice-fish farming system are 4184 kg and 1233 kg per acre respectively. In term of return on production, mixed rice-fish producer gets on an average BDT 152900.7 and mono rice producer obtains BDT 96574.25 per annum. Results from regression models provide statistical evidence that mixed rice-fish farming is more efficient than mono rice farming in the area. Estimated coefficients of the model also show that mixed farming enjoys increasing returns to scale, whereas, mono rice farming system is being operated under decreasing returns to scale. Test of hypothesis also supports the notion that mixed rice-fish farming system is more profitable than mono rice farming. It is important for the concerned policy maker to take steps for promoting mixed farming to remove poverty and ensure sustainable rural development in Bangladesh.

REFERENCES

1. Ahmed N, and Garnett ST, 2011. Integrated rice-fish farming in Bangladesh: meeting the challenges of food security. Food Security, 3: 81-92.
2. Ahmed N, Zander KK, and Garnett ST, 2011. Socioeconomic aspects of rice-fish farming in Bangladesh: opportunities, challenges and production efficiency. Australian Journal of Agricultural and Resource Economics, 55: 199-219.
3. Asiatic Society of Bangladesh, 2006. Banglapedia, Asiatic Society of Bangladesh, Dhaka.
4. BBS, 2013. Bangladesh Economic Review 2013, Ministry of Finance, Dhaka, Government of The People's Republic of Bangladesh.
5. Dey MM, Spielman DJ, Haque AM, Rahman M, and Valmonte-Santos R, 2013. Change and diversity in small holder rice–fish systems: Recent evidence and policy lessons from Bangladesh. Food Policy, 43: 108-117.
6. Hasanuzzaman A, Rahman M, and Islam S, 2011. Practice and economics of freshwater prawn farming in seasonally saline rice field in Bangladesh. Mesopotamian Journal of Marine Science, 26: 69-78.
7. Mamun SA, Nasrat F, and Debi MR, 2012. Integrated farming system: prospects in Bangladesh. Journal of Environmental Science and Natural Resources, 4: 127-136.
8. Nahar A, 2010. Impacts of different rice-fish-prawn culture systems on yield of rice, fish and prawn and limnological conditions. Journal of the Bangladesh Agricultural University, 8: 179-185.
9. Noorhosseini-Niyaki SA, and Bagherzadeh-Lakani F, 2011. 'Ecological and biological effects of fish farming in rice fields', Regional Congress of Sustainable Management Science-based in Agriculture and Natural Resources. Gorgan University of Agricultural Sciences and Natural Resources, Iran, pp. 21-22.
10. Roy AK, Ghosh AK, Arafat ST, and Huq KA, 2013. Fish-paddy crop rotation practice in South-West coastal region of Bangladesh: A profitable technology for the poor farmers. Bangladesh Journal of Agricultural Research, 38: 389-399.

15

EFFECT OF STOCKING DENSITY OF FINGERLINGS PRODUCTION OF BLACKCARP *Mylopharyngodon piceus* (J. Richardson, 1846) IN POND CONDITION

Md. Abdus Samad[2*], Mst. Lutfunnahar[2], Sujit Kumar Chatterjee,[3]Md. Ashrafuzzaman and [4]Md. Selim Reza[2]

[2]Department of Fisheries, University of Rajshahi, Rajshahi-6205, Bangladesh; [3]Fish inspection and quality control officer, DOF, Dhaka, Bangladesh; [4]SUFO, DOF, Panchagarh Sadar, Bangladesh

*Corresponding author: Md. Abdus Samad; E-mail: samad1413@yahoo.com

ARTICLE INFO	ABSTRACT

Key words

Stoking density
Growth
Production
Black carp

An experiment was conducted in pond condition under three treatments for a period of 60 days each with three replications to know the effect of stocking density on growth performance of fingerlings production of Black carp *Mylopharyngodon piceus*. Stocking density were used at the rate of 240 (T_1), 270 (T_2) and 300 (T_3) fingerlings/decimal, respectively. Initial average length and weight of *M. piceus* fry were 3±0.02 cm and 1.5±0.01g during stocking in the experimental ponds. The fish were initially fed with 25% protein content formulated feed at 8% of body weight and the rate was reduced to 6% gradually. The water quality variabilities were more or less similar in three treatments within the suitable ranges for aquaculture. The SGR value 3.57±0.006 was recorded in treatment T_1 while the lowest value was 3.20±0.009 in T_3. Survival rate (%) were significantly higher in T_1 (91.33±0.88) where the stocking density was low compared to those in T_2 (79.83±0.44) and T_3 (77.17±0.6), respectively. The net profit in treatment T_1 was (Tk. 475428.58±3.25) and lowest in T_3 (Tk. /ha 190138.5±6.15). The highest final weight gain, SGR%, production, net profit and cost benefit ratio were found in treatment T_1. In the present study production of *M. piceus* was found to be highest in lower stocking density. Therefore, it is evident that feeding with higher protein supplement with lower stocking density is effective for optimum growth of *M. piceus*.

INTRODUCTION

Aquaculture is the cultivation of fish in captivity under controlled conditions. At present it is an important and rapidly expanding enterprise all over the world. In fish farming the management is directed towards the realization of maximum fish biomass within a certain time (Verreth, 1999; De-Franceso *et al.,* 2007; Schuchardt *et al.,* 2008). In Indian subcontinent, fisheries has always played a pivotal role in the food and nutritional security of the people (Sugunan, 2002). Black carp in Bangladesh is also called snail carp (Rahman, 2005) as it can reduce and control snail's biomass as well as increase the carrying capacity of the pond. Fingerling of Black carp is not available in Bangladesh for grow-out production. Therefore, a suitable culture method for rearing of *M. piceus* fry is very important to ensure reliable and regular supply of fingerlings.

Changes of population densities of fishes may lead to changes in growth and survival rate (Miao, 1992). Fish larvae grow slowly and have a low survival rate at high stocking density (Huang and Chiu, 1997). The density at which a fish species can be stocked is an important factor in determining the economic viability of a production system in intensive aquaculture (Papst *et al.,* 1992).The high stocking density, however, may exert adverse effects on growth and survival. Therefore, it is necessary to predetermine and standardize the optimum stocking density for each species in order to obtain the best possible output.

MATERIALS AND METHODS

Research area and time

The experiment was carried out for a period of 60 days from 17th July to 16th September, 2013 in Rajshahi University hatchery complex. Initial average length and weight of *M. piceus* fry 3.0±0.02 cm in length and 1.5±0.01g in weight were stocked in the experimental ponds. The experimental ponds were rectangular in shape with similar size, depth and bottom type including water supply facilities. The average area of the ponds was 0.60 decimal (0.0024 hector). The water depth was maintained around 1.0-1.25 m.

Experimental design

The experiment was conducted in the pond into three treatments viz. T_1, T_2 and T_3. Each experiment having three replications for evaluation of fingerlings production of *M. piceus*. In this experiment stocking density were 240/decimal (59280/ ha), 270/decimal (66690/ha) and 300/decimal (74100/ha) in the treatment T_1, T_2, T_3 respectively.

Pond preparation

Rotenone (19.76 kg/ha) was used for the removal of predatory and unwanted fish species. After 7 days, all ponds were treated with lime at the rate of 247 kg/ha to disinfect the water. The sources of water of experimental ponds were rainfall and deep tube-well. During the introduction of water in each experimental pond, fine mesh (2 mm) nylon net hapa was used in the mouth of the pumped water to prevent predatory fish egg, spawns, fry and adult or fry of aquatic harmful insects to inhabit their entrance.

Collection and stocking of Fry

Fry of black carp were collected from private hatchery of Jessore. Before releasing the fry to the experimental pond the initial length and weight of 10 fry were recorded with the help of measuring scale and a sensitive portable electric balance (KD300kc: 0.01g-300g).

Feed preparation and feeding

The supplemental feed was given to fry at the rate of 8%, 6% in 1st and 2nd month respectively. The quantities of feed were adjusted every 15 days interval on the basis of increase in the average body weight of the stocked biomass. Half of the ration was supplied at 9.00 am and remaining half was supplied at 4.00 pm. The proximate composition of feed has been presented in the table 1 and 2.

Table 1. Composition of feed used in the experiment

Ingredients	Inclusion rate (%)
Fish meal	20
Rice bran	30
Wheat flour	30
Mustard oil cake	20

Table 2. Proximate composition of feed used in the experiment

Components	Diets
Moisture	13.0
Crude protein	25.0
Crude lipid	10.98
Crude Fiber	11.34
Ash	10.2
NFE	29.48

* Nitrogen free extract (NFE) calculated as 100% (Moisture + Crude protein+ Crude lipid+ Crude Fiber+ Ash)

Sampling procedure

Sampling was done on every fortnight interval by random sampling of 10% fry from each experimental pond by using a small net. Weight was taken with an electric balance and length was recorded with measuring board. All the collected data were recorded and finally calculated the average length and weight of fry according to treatment on each sampling day.

Water quality parameters

Physico-chemical parameters like water temperature (C), transparency (cm), Dissolved oxygen (mg/1), CO_2 (mg/l), NH_3-N (mg/1), pH, Alkalinity (mg/1) of each experimental pond were measured at 15 days interval and data were recorded. Temperature was recorded by using a Celsius thermometer, transparency was recorded by secchi disc and other chemical parameters were recorded by using Hack kit box (HACK kit, FF-2, USA).

Growth parameters

The growth, length in cm and weight in g was measured in every 15 days interval. To evaluate the fish growth the following parameters were measured.

i. Weight gain (g) = Average final weight - Average initial weight

ii. Length gain (cm) = Average final length - Average initial length

iii. Specific growth rate (SGR) = $\dfrac{LnW_2 - LnW_1}{T_2 - T_1} \times 100$

Where, W_2 = Final live body weight (mg) at time T_2
W_1 = Initial live body weight (mg) at time T_1

iv. Survival rate was calculated by the following formula =

$\dfrac{\text{Initial number of fry} - \text{Final number of fry}}{\text{Initial number of fry}} \times 100$ (Brown, 1957)

v. Production of fishes = No. of fish harvested x final weight of fish.

Economics analysis

A simple economic analysis was done to estimate the economic return in each treatment for experiment. The total cost of inputs was calculated and the economic return was determined by the differences between the total return (from the current market prices) and the total input cost.

Statistical Analysis

For the statistical analysis of data collected, one-way analysis of variance (ANOVA) was performed using the SPSS (Statistical Package for Social Science, evaluation version-15.0). Significance was assigned at the 0.05% level. The mean values were also compared to see the significant difference through DMRT (Duncan Multiple Range Test) after Zar (1984).

RESULTS

Water quality parameters

Water quality parameters were monitored fortnightly. Water temperature, water transparency, dissolved oxygen, free carbon dioxide, pH, ammonia-nitrogen and total alkalinity of water varied from 29.12 ± 0.02 to $32.02\pm0.0.04°C$, 24.70 ± 0.44 to 34.30 ± 0.4 cm, 4.40 ± 0.21 to 5.46 ± 0.17 mg/1, 2.69 ± 0.21 to 3.39 ± 0.03 mg/1, 7.61 ± 0.02 to 7.97 ± 0.24, $0.09\pm0.0.01$ to 0.14 ± 0.003 mg/1and 89.78 ± 0.89 to 107.0 ± 5.19 mg/1respectively. Water temperature, water transparency, dissolved oxygen, free carbon dioxide, pH, ammonia-nitrogen and total alkalinity of water were not significantly ($p<0.05$) different among the parameters. Variation in the mean values of water quality parameters in three different treatments were showed in table 03.

Table 3. Variations in mean values of physico-chemical characteristics under different treatments during the study period

Parameters	Treatments		
	T_1	T_2	T_3
Temperature (°C)	30.69 ± 0.61^a	30.55 ± 0.54^a	30.61 ± 0.61^a
Transparency (cm)	29.38 ± 1.67^a	28.86 ± 1.41^a	30.22 ± 0.69^a
DO (mg/l)	4.86 ± 0.21^a	5.07 ± 0.14^a	5.11 ± 0.16^a
CO_2 (mg/l)	3.08 ± 0.14^a	2.98 ± 0.07^a	3.31 ± 0.21^a
pH	7.76 ± 0.03^a	7.82 ± 0.07^a	7.72 ± 0.04^a
Alkalinity	99.46 ± 4.00^a	97.86 ± 3.33^a	94.20 ± 2.09^a
NH_3-N	0.11 ± 0.02^a	0.13 ± 0.01^a	0.13 ± 0.01^a

Figures in a row bearing common letter do not differ significantly ($p<0.05$)

Growth performance of fish

In the present experiment there was no significant ($P<0.05$) difference in initial weight of fish under different treatments. The average final weights were 12.74 ± 0.04g, 11.18 ± 0.08g and 10.22 ± 0.06g in T_1, T_2 and T_3 respectively. Weight increments were statistically significant among the treatments. The highest growths in weight were observed in T_1 (12.74 ± 0.04) and lowest in T_3 (10.22 ± 0.06). Similarly, the mean length gain in treatments T_1, T_2 and T_3 were 11.03 ± 0.04, 9.34 ± 0.04 and 8.23 ± 0.05 cm respectively. The highest growths in length were noticed in T_1 (14.03 ± 0.04cm) and the lowest in T_3 (11.23 ± 0.05cm). The recorded mean specific growth rate after 60 days of experiment of treatments T_1, T_2 and T_3 were 3.57 ± 0.06, 3.35 ± 0.01 and 3.20 ± 0.009 (table 4) respectively, which were significantly ($P<0.05$) different among the treatments. The significantly ($P<0.05$) highest SGR ($\%$, bwd^{-1}) value 3.57 ± 0.006 was recorded treatment T_1 while the lowest 3.20 ± 0.009 was obtained in T_3. The survival rate ($\%$) in different treatments was fairly high. The survival ranged between 77.17 ± 0.66 to 91.33 ± 0.33. Significant ($P<0.05$) different survival rate were found among the treatments. The production of *M. piceus* ranged between 1286.6 ± 100.19 (T_3) to 1690.8 ± 106.5 (T_1) kg/ha/60days in different treatments.

Economic analysis

A simple economic analysis was performed to estimate the net profit from this culture operation. The cost of production was based on the local wholesale market price of the input used of the year 2013.The cost of leasing ponds was not included in the total cost. The cost of different inputs and economic return from the sale of fishes in different treatments are summarized in table 05. The total cost of inputs and profit per hectare were significantly different (P<0.05) among the treatments. The cost input was lowest in T_1 treatment followed by T_2 and T_3. The net profit was highest in T_1 treatment and lowest in T_3 treatment which was significantly different among the treatments, cost and benefit ratio were calculated 1:1.41, 1:0.84, 1:0.48 among T_1, T_2 and T_3 respectively.

Table 4. Growth and production performance of *M. piceus* under different treatments during the study period

Growth parameters	Treatments		
	T_1	T_2	T_3
Weight gain (g)	11.24±0.04[a]	9.68±0.08[b]	8.1±0.06[c]
Length gain (cm)	11.03±0.04[a]	9.34±0.04[b]	8.23±0.05[c]
SGR(%,bwd^{-1})	3.57±0.006[a]	3.35±0.01[b]	3.20±0.009[c]
Final weight (g)	12.74±0.04[a]	11.18±0.08[b]	10.22±0.06[c]
Final length (cm)	14.03±0.04[a]	12.34±0.04[b]	11.23±0.05[c]
Survival rate (%)	91.33±0.88[a]	79.83±0.44[b]	77.17±0.6[c]
Yield (kg/ha/60 days)	1690.8±106.5[a]	1413.7±181.84[b]	1286.6±100.19[c]

Figures in a row bearing common letter(s) do not differ significantly (p<0.05)

Table 5. Input cost and profit from *M. piceus* for 60 days in ponds of three different treatments

Components	Treatments		
	T_1	T_2	T_3
Initial pond preparation (Tk/ha)	32110.4±00[a]	32110.4±00[a]	32110.4±00[a]
Fry cost (Tk/ha)	237112±00[c]	266760±00[b]	296400±00 [a]
Feed cost (Tk/ha)	56285.4±2.48[c]	57422.4±1.03[b]	58529.4±3.26 [b]
Operational cost (Tk/ha)	10150±00[a]	10150±00[a]	10150±00[a]
Total cost (Tk/ha)	336155±32.04[c]	366250±14.15[b]	397387±8.45[a]
Total income (Tk/ha)	811584±3.28[a]	672075±3.32[b]	587835.5±6.15 [a]
Net profit (Tk/ha)	475428.58±3.25[a]	305819.8±3.32[b]	190138.5±6.15[c]
CBR	1.41±0.01[a]	0.84±0.009[b]	0.48±0.02[c]

Figures in a row bearing common letter(s) do not differ significantly (p<0.05), *Leasing cost is not included

DISCUSSION

Water quality parameters

The mean values of secchi disk depth (cm) among the treatments were 28.86±1.41, 29.38±1.67 and 30.22±0.69 cm in treatments T_1, T_2 and T_3 respectively. Consistently higher transparency depth was recorded in T_3, which might be due to the reduction of the plankton population by higher density of fish (Haque *et al.*, 1993; 1994). Boyd (1990) recommended a transparency between 15 cm to 40 cm as appropriate for fish culture. Wahab *et al.* (1995) suggested that the transparency of productive water should be 40 cm or less. The mean values of water temperature in treatment T_1, T_2 and T_3 were 30.69±0.61°C, 30.55±0.54°C, 30.61±0.61°C respectively Quddus and Banerjee (1989) denoted that the water temperature between 29°C and 32°C is suitable for the faster growth of fish spawn and aquatic organisms under natural conditions. Rahman *et al.* (1992) also found water temperature ranged 25.5°C to 30.0°C, which is favorable for fish culture. The range of temperature (30.55–30.69°C) in the experimental ponds were within the acceptable range for nursing of fry and fingerlings of warm water fishes that agreed well with the findings of Haque *et al.* (1993; 1994), Kohinoor *et al.*

(2001), Rahman (2005) and Britz and Hecht (1987).In the present study the DO content in water was 4.86±0.21 to 5.11±0.16 mg/l. More or less similar results were reported by Hossain (2000) and Kohinoor et al (2001) study two ponds where they recorded DO values of fish ponds ranged from 3.8 to 6.9 mg/L and 2.04 to 7.5 mg/L respectively.

According to Rahman (1992) DO content of a productive pond should be 5 mg/l or more. The mean values of pH in treatments T_1, T_2 and T_3 were 7.76±0.03, 7.82±0.07 and 7.72±0.04 respectively. The pH values of the present study are also agreed with the findings of Hossain et al., (2013), Chakraborty and Mirza (2007), Kohinoor et al. (1994)

In the present study mean values of total alkalinity were 99.46±4.00, 97.86±3.33 and 94.20±2.09 mg l^{-1} in treatment T_1, T_2 and T_3, respectively. The findings of the present study are in agreement with those of Islam (2002), Rahman and Rahman (2003), and Rahman et al. (2004, 2005). The mean values of ammonia-nitrogen were 0.11±0.02, 0.13±0.01 and 0.13±0.01 mg l^{-1} in T_1, T_2 and T_3 respectively. Kohinoor et al. (2001) recorded ammonia-nitrogen ranging from 0.01 to 1.55 mg/L in monoculture ponds. However, the present level of ammonia-nitrogen content in the experimental ponds is not lethal to the fish (Kohinoor et al., 2001). So, in the present study ammonia-nitrogen value was suitable for fingerling rearing. The mean value of CO_2 of water varied from 2.98±0.07 to 3.31±0.21 in T_1, T_2 and T_3 treatment respectively. DoF (2008) found that free CO_2 level of 1.04-29.49 mg/l which were more or less similar to the present study.

Growth and production performances

Growth in terms of final length, length gain, final weight gain and specific growth rate of fingerlings of M. piceus were significantly higher in T_1 where stocking density of fingerlings is 59280/ ha was low compared to these of T_2 66690/ha and T_3 (74100/ha) although the same food was applied at on equal ratio in all the treatment. Stocking density had previously been observed to have a direct effect on growth of fish (Islam 2002; Rahman et al., 2004). High stocking density of larvae in combination with abundant food in the rearing system might produce a stressful situation if not from the build-up of metabolites than from competitive interaction (Haque et al., 1994). Significantly higher survival rate of fingerlings was obtained in T_1 (91.33±0.88), where the stocking density was low compared to those in T_2 (79.83±0.44) and T_3 (77.17±0.6). The causes from decreasing survival rates in those treatments were well as competition for food and space in the experimental ponds. This agrees well in the findings of Kohinoor et al., (1994) and Rahman et al., (2005) during fry rearing experiments of various carp species.

In the present study, significantly higher net production of fingerling were obtained from ponds stocked with 59280 fingerlings/ha than those from the ponds stocked with 66690 fingerling/ha and 74100/ha, including that the growth and percentage of survival decreased with increasing stocking density. The results in the present experiment are very close to those of Saha et al. (1988) who obtained a gross production of 1385.15 to 1995.60 kg/ha by 8 weeks rearing of Ruhu (Labeo rohita) fingerlings at 60000 to 80000/ ha stocking density. Similar production of 1869.1 kg/ha^{-1} by rearing of Labeo calbasu fingerlings for 8 weeks at a stocking density of 80000 fingerlings/ha. Significantly high numbers of fingerlings were produced in T_1 where the stocking density was lower than those in T_2 and T_3. The higher market price of larger fingerlings produced in ponds with 59280 fingerling/ha substantially increased the net benefit compared to those obtained from the smaller fingerlings produced at higher stocking density

Among the treatments, the highest production was also found in T_1 and consequently provides the highest net profit (Tk.475428.58±3.25) with $T_1$where fishes were stocked at 59280/ha. Similarly, the net profit in treatment T_2 was (Tk. 30581908±3.32) and lowest in T_3 (Tk./ha 190138.5±6.15).Cost benefit ratio(1.41±0.01) was also higher in T_1 which was significantly different among the treatment. . The finding is more or less similar to the finding with Bob Manuel and E.S, Erondu (2010) who found CBR of nile tilapia O. niloticusas 1.60-2.03 and Ali et al. (2011) found CBR as 2.60 Tk. Overall, the highest growth, survival, production and benefit of fingerlings were obtained in ponds with 59280 fingerlings/ha compared to the ponds with higher stocking densities.

ACKNOWLEDGEMENTS

We express our gratitude to Department of Fisheries, Rajshahi University, Bangladesh for the facilities provided to carry out this research work.

REFERENCES

1. Ali MS, MA Hossain, and MN Naser, 2011. Species suitability for small scale cage aquaculture in river ecosystem of northern Bangladesh. Bangladesh Journal of Progressive Science and Technology, 9: 197-200.

2. Banerjee SM, 1967. Water quality and soil conditions of fishponds in some states at India in relation to fish production. Indian Journal of Fisheries, 14: 115-155

3. Bhuyan B R, 1970.Physico-chemical qualities of water of some Ancient tanks in Sibsagar, Assam. Environmental Health, 12: 129-134

4. Bob-Manuel FG and ES Erondu, 2010. Yeast single cell protein in the diet of *Oreochromis niloticus* (L) fingerlings: An economic evaluation. African Journal of Biotechnology, 10: 18581-18584

5. Boyd CE, 1990. Water Quality in Ponds for Aquaculture. Birmingham Publishing Co. Birmingham, Alabama, USA. 477 pp.

6. Britz P J and T Hechet, 1987. Temperature preferences and optimum temperature for growth of African catfish (*Clarias gariepinus*) larvae and post larvae. Aquaculture, 63: 205-214

7. Brown M E, 1957. Experimental studies on growth. Vol.1, Academic press, New York, 361-400 pp.

8. Chakraborty BK and MJA Mirza, 2007. Effect of stocking density on survival and growth of endangered bata, *Labeo bata* (Hamilton–Buchanan) in nursery ponds. Aquaculture, 265: 156-162.

9. DoF, 2008. Matsha Pakkah Shankalan, Department of Fisheries, Ministry of Fisheries and Livestock, Government of the People's Republic of Bangladesh, pp. 79-81.

10. Haque MZ, Rahman MA and M.M Hossain, 1993. Studies on the effect of stocking densities on the growth and survival of mrigal (*Cirrhinus mrigala*) fry in rearing ponds. Bangladesh Journal of Zoology, 21: 51-58.

11. Haque MZ, Rahma MA., Hossain, MM and MA Rahman, 1994. Effect of stocking densities on the growth and survival of mirror carp, *Cyprinus cerpio* var. *Specularis* in rearing ponds. Bangladesh Journal of Zoology, 22: 109-116.

12. Hossain MY, 2000. Effects of iso-phosphorus organic and inorganic fertilizer on water quality parameters and biological production. M.S. Thesis, Department of Fisheries Management, BAU, Mymensingh-2202, pp 74.

13. Hossain MI, S Ahmed, MS Reza, MY Hossain, M Islam, A Jesmin and R Islam, 2013. Effects of organic fertilizer and supplementary feeds on growth performance of silver carp (*Hypophthalmichthys molitrix*) and bata (*Cirrhinus reba*) fry in nursery ponds. International Journal of Research in Applied, Natural and Social Sciences (IJRANSS), 1: 117-124.

14. Huang WB and TS Chiu, 1997. Effects of stocking density on survival, growth, size variation, and production of Tilapia fry. Aquatic Research, 28: 165-173.

15. Islam MS 2002. Evaluation of supplementary feeds for semi-intensive pond culture of mahseer, *Tor putitora* (Hamilton). Aquaculture, 212: 263-276.

16. Jhingran VG and RSV Pullin, 1985. A Hatchery Manual for the Common Carpu, Chinese and Indian Majors Carps. ICLARM Studies and Reviews, vol. II. 191 pp.

17. Kohinoor AHM, MZ Haque, MG Hussain and MV Gupta, 1994. Growth and survival of Thai punti, *Puntius gonionotus* (Bleeker) spawn in nursery ponds at different stocking densities. Journal of Asiatic Society of Bangladesh, Science, 20: 65–72.

18. Kohinoor AHM, Wahab MA, Islam ML and SH Thilsted, 2001. Culture potentials of mola (*Amblypharyngodon mola*), chela (*Chela cachius*) and punti (*Puntius sophore*) under monoculture system. Bangladesh Journal of. Fisheries Research, 5: 123-134.

19. Pandian TJ, SN Mohanty and S Ayyappan, 2001. Food requirements of Fish and Feed Production in India. In sustainable Indian Fisheries (Pandian, T.J. Ed). Nation Academy of Agriculture, Science, New Delhi, pp 153-165.

20. Papst MH, TA Dick, N Arnason and CE Engel, 1992. Effect of rearing density on the early growth and variation in growth of juvenile Arctic charr, *Salvelinus alpinus* (L.). Aquaculture Research, 23: 41-47.

21. Quddus MMA and AK Banerjee, 1989. Diurnal variations in the physico-chemical parameter of nursery pond. Bangladesh Journal of Aquaculture, 11-13: 47-51.

22. Rahman AKA, 2005. Freshwater fishes of Bangladesh, 2nd ed. Zoological Society of Bangladesh, Department of Zoology, University of Dhaka, Dhaka, 394 pp.

23. Rahman MS, 1992. Water Quality Management in Aquaculture BRAC Prakashana. 66, Mohakhali, Dhaka –1212, Bangladesh, 75pp.

24. Rahman MA and MR Rahman, 2003. Studies on the growth and survival of sharpunti (*Puntius sarana* Ham.) spawn at different stocking densities in single stage nursing. Progressive Agriculture, 14: 109-116.

25. Rahman MR, Rahman, MA and MG Hussain, 2004. Effects of stocking densities on growth, survival and production of calbasu (*Labeo calbasu* Ham.) in secondary nursing. The Bangladesh Veterinarian, 21: 58-65.

26. Sah SB, MV Gupta, MG Hussain and MS Shah, 1988. Growth and survival of rohu (*Labeo rohita* Ham.) fry in rearing ponds at different supplementary feeding. Bangladesh Journal of Zoology, 16: 119-126.

27. Schuchardt D, JM Vergara, HF Palaciso, CT Kalinowski, CMH Cruz, MS Izquierdo and L Robaina, 2008. Effects of different dietary protein and lipid levels on growth, feed utilization and body composition of red porgy (*Pagrus sp.*) fingerlings. Aquaculture Nutrition, 14: 1-9.

28. Steven C and LA Helfrich, 2002. Understanding fish nutrition, feeds, feeding. 420-456pp.

29. Sugunan VV, 2002. Enhancement: an effective tool for increasing inland fish production. Fishing Chimes, 21: 21-22.

30. Tahir M ZI, A Iftikhar, A Mateen, M Ashaf, ZH Naqvi and H Ali, 2008. Studies on partial replacement of fish meal with oilseeds meal in the diet of major carps. International Journal of Agriculture & Biology, 10: 455-458

31. Wahab MA, ZF Ahmed, MA Islam and SM Rahmatullah, 1995. Effect of introduction of economic carp *Cyprinus carpio* (L) on the pond ecology and growth of fish in polyculture. Aquaculture Research, 26: 619-628.

32. Wurts WA, 2005. Review of feeding practices for channel catfish production. Kuntucky State University CEP at UK Research and Education Center, P.O. Box. 469.

33. Zar JH, 1984.Biostatistical analysis. 2nd edition. Englewood Cliffs, NJ: Prentice- Hall. 130.

16

SHORE TO LANDWARD TRANSECT BURROW DIVERSITY OF FIDDLER CRAB IN A TROPICAL INTERTIDAL COAST OF CHITTAGONG IN BANGLADESH

Mst. Mansura Khan[1*] and Mohammad Sadequr Rahman Khan[2]

[1]Department of Fisheries Biology and Genetics, Faculty of Fisheries, Bangladesh Agricultural University, Mymensingh-2202, Bangladesh; [2]Department of Marine Bioresources Science, Chittagong Veterinary and Animal Sciences University, Chittagong-4225, Bangladesh

*Corresponding author: Mst. Mansura Khan; E-mail: keya.3a@gmail.com

ARTICLE INFO

Key words

Burrow characters
Saltmarsh
Mangrove
Mangrove pool

ABSTRACT

Burrows indicate the abundance and distribution of fiddler crab in an intertidal coast that varies with structure and morphology within intertidal habitats. We observed fiddler crab burrow density and characters (burrow length, depth, diameter and volume) within randomly selected six 1m² quadrate from three intertidal habitats: higher saltmarsh, mangrove pool (a small ditch distributed within mangrove) and mangrove land through field surveys in a coast of Chittagong, Bangladesh. Burrows were observed and counted for density estimation, and burrow characteristics were studied through excavating 10 representative burrows from each quadrate of each habitat. Spearman correlation was used to relate between the distances (from shore towards land) and burrow characters. Transect starting from saltmarsh as base towards mangrove land showed burrow density decreased from shore to higher intertidal habitat. Simultaneously, higher burrow length and diameter were observed landward and contrariwise shoreward. Burrow prevalence in mangrove pools represents fiddler crabs are abundant within land and shore interface presumably due to the dual privilege of easy burrowing and moist condition required for gill ventilation.

INTRODUCTION

Fiddler crabs of the genus *Uca* Leach, 1814 are variably distributed over tropical to temperate mudflats, salt marshes, mangrove swamps, and sandy beaches (Crane, 1975; Hodgson, 1987; Peer, 2015). This genus is noteworthy with sexual dimorphism with male having one astoundingly enlarged and one minor cheliped, whereas females possess two equally sized chelipeds (Peer, 2015). About 97 species of *Uca* were recorded in the tropical and temperate climates throughout the world (Rosenberg, 2001) and by and large they live in the intertidal zone, actively create burrows during low tide as they stay inside their burrows during high tide (Wolfrath, 1992). Burrowing begins when crabs are very small (Herrnkind, 1968) and is influenced by a variety of conditions, such as water availability, food availability, substratum, ground temperature, tidal periodicity, reproductive activity, threat by potential predators, seasons and mate activities (Bertness, 1985; Genoni, 1991).

Burrows are exigent to fiddler crabs that support to adapt in a semi-terrestrial life to get rid of excessive wave action and environmental stresses to relief from harsh ambient temperatures and desiccation (Bertness and Miller, 1984; Lim and Diong, 2003). Burrows provide a shelter and escape route from terrestrial predators during exposed at low tide and from aquatic predators during at high tide, water for physiological processes, sites for molting and reproductive needs (Hyatt and Salmon, 1979; Christy, 1982; Thurman, 1984; Christy, Backwell *et al.*, 2001; and Milner *et al.*, 2010).

Fiddler crab creates burrows in intertidal zone with patchiness in distribution and diversity in burrow morphology. Within intertidal zone, salt marsh and mangrove habitats are the most productive ecosystems in tropical and subtropical regions lying between land and sea (Chowdhury *et al.*, 2011) and these habitats provide shelters for fiddler crabs. By means of burrow creation, fiddler crab work for nutrient recycling and energy flow in coastal ecosystem (Qureshi and Saher, 2012), translocate oxygen into the anoxic layers and promote aerobic respiration, iron reduction and nitrification thus play an important role in biogeochemical cycle of mangrove sediments (Mokhtari *et al.*, 2016). Likewise, they convert organic matter into small sized packs for others organisms and being preyed upon by predators and transfer energy both in terrestrial and marine environment (Skov and Hartnoll, 2001; Tina *et al.*, 2015).

Spatial distribution of fiddler crabs was studied in different coast of the world depends on the availability of vegetation, substratum, food, salinity, tide, and presence of other animals (Icely and Jones, 1978; Nobbs, 2003; César *et al.*, 2005). Despite of huge ecological significance in intertidal habitat and large coastal region, research regarding spatial distribution, burrow characteristics of fiddler crab are scanty and infancy in Bangladesh. Therefore, with a view to understanding burrow distribution pattern in intertidal habitats of Bangladesh, we surveyed a coast of Chittagong for the first time, which will be an effigy of other coasts and will open a horizon for future research.

MATERIALS AND METHODS

Sampling area

Sampling took place in a tropical intertidal coast during low tide at South Kattoly sea beach (Figure 01), locally known as Dakshin Kattoly sea beach, Salimpur, Chittagong (Latitude $22^0 35'50"$ N, Longitude $91^0 75'40"$ E) along the Bay of Bengal. This coast was planted by mangrove forest under the Asian Development Bank green belt project on 1997-1998 and characterized by a long intertidal salt marsh merged with mudflat and mangroves stretches about 200 m from the shore line. Tides are semi-diurnal with high tide and low tides in every 6 hours interval.

The characteristic of the coast is muddy or sandy coast dominated by saltmarsh *Porteresia coarctata* (Aysha *et al.*, 2015), and mangrove was dominated by *Avicennia spp.* Fiddler crab was found in higher marsh but not in lower saltmarsh. A special feature of around 1ft deep irregular shaped ditch (round or oval or L-shaped,), mangrove pool was observed within mangrove land adjacent to salt marsh (Fig. 3A). Pools remain water logged as it receives water during spring tide, but usually get dry when normal tidal water cannot reach.

Sampling method and data collection:

Sampling was followed by transect from saltmarsh towards mangrove by selecting three habitats (saltmarsh, mangrove pool and mangrove) and data were collected from November 2015 to February 2016. As burrow count provide quick and reliable estimate (Warren, 1990), burrow abundance and burrow characteristics (diameter, length, depth, volume and shapes) were determined in three intertidal habitats from shore towards land viz. higher saltmarsh, mangrove pools and mangrove higher intertidal land. In each habitat, we selected six 1 m^2 (1 m ×1 m) quadrates randomly to count burrows visually. But three quadrates in saltmarsh were chosen intentionally along the same line to consider them as base point to measure landward distance of other quadrates. Burrow length, depth and shape were measured with measuring tape after excavating 10 representative burrows from each quadrate using soil corer up to 30 cm depth. Burrow diameter was measured by digital venire calipers. Burrow volume was calculated using the formula ($\pi r^2 h$, where π = 3.1416, r = radius, h = length) assuming a cylindrical hole. The distance of each quadrate was measured from the first three quadrates which were situated in saltmarsh. In order to find correlations between distances (shores to landward) and burrow characters, each quadrate distance was measured from the quadrate positioned in salt marsh considering as the base point (0 m) towards mangrove, so that we can understand sea to landward burrow characters.

Data Analysis

Data were checked for homogeneity with Levine's test and for normality with Shapiro-Wilk test. We compared means of burrow characters among three habitats using one-way ANOVA through SPSS version-16.0. Significance was assigned at the 0.05% level. Bivariate Pearson correlation analysis (2-tailed) was used to find correlation between shore to landward distance and burrow characters.

RESULTS

Fiddler crab burrows were diversely distributed among three habitats in the intertidal zone, and the burrow characters differed considerably (Table 1). Considering burrow density, significantly higher (P<0.05) number of burrows (83.67±16.33) was found in mangrove pool and the highest abundance was 115 (Table 1). Within other habitats, relatively higher amount of burrows were observed in saltmarsh areas (76.67±6.67) than in mangrove intertidal high land (42.00±3.79). And, the highest number of burrows was 90 and 49 in those two habitats, respectively.

There was no significant difference among the shelters regarding burrow length and depth. Minimum and maximum length was 26.0 and 58.2 mm in saltmarsh areas, 44.80 and 89.60 mm in mangrove pool, and 55.0 and 112.0 mm in mangrove high land. In these sites, average burrow length was 43.4±9.4, 69.5±13.1, and 90.0±17.7 mm, respectively (Table 1). Burrow length increased with increasing closeness to high intertidal land, and reciprocal observation in seawards boundary towards saltmarsh (Figure 3). Likewise, depths of the burrows were recorded lowest 38.0±8.1 mm in saltmarsh, highest 76.7±13.6 mm in mangrove land and 60±10.1 mm in mangrove pool. Similarly, burrow diameter was significantly higher (P<0.01) in mangrove land (24.0±2.1 mm) and mangrove pool (18.9±2.0 mm) than in higher saltmarsh (12.3±0.3 mm). Therefore, burrow lengths, depths and diameters were higher in mangrove higher intertidal land than the other habitats.

Burrows were also observed adjacent to a 2-5 feet water channel with patchy distribution that remains dry in low tide and drain tidal salt water upward during spring tide or downward during low tide. Four distinct shapes of burrows (I, J, C and L shaped) were observed with predominant 'I' shaped burrows in saltmarsh and, I, J and C shaped burrows in mangrove pool and mangrove higher intertidal habitat (Figure 4). Burrows with higher length had the enlarged ground in the bottom (Figure 4B and 4C).

Correlation

A Pearson product-moment correlation coefficient was computed to assess the relationship between shore to landward distances and burrow characteristics. Burrow lengths and burrow diameters were significantly positively correlated with landward distance (r=0.774, p=0.014, and r= 0.811, p= 0.008, respectively). Contrarily, negative relation was found between distance and burrow length (Table 2). Both the burrow length and burrow diameter were negatively related with burrow numbers.

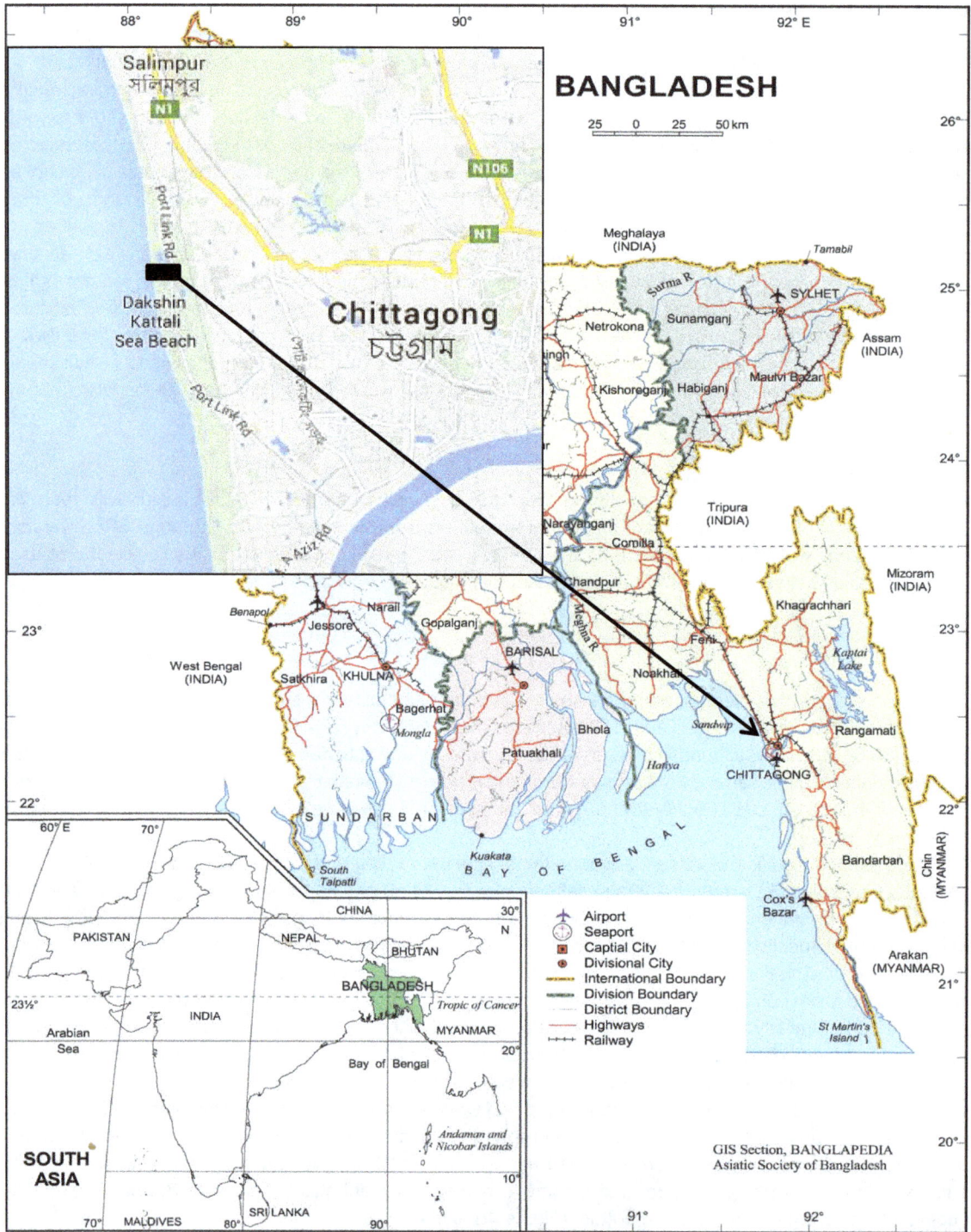

Figure 1. Geographical location of the sampling site with map from Google earth and Asiatic Society of Bangladesh

Table 1. Burrow characteristics (Mean ± SE) in salt marsh, mangrove pool and mangrove land. The range of observed values is given in the parentheses. Means with the superscripts are significantly different (P<0.05) based on Duncan's test. N=6 quadrates (10 burrows from each)

Characters	Saltmarsh	Mangrove pool	Mangrove land	p-value
Number of burrows	76.67±6.67[ab]	83.67±16.33[a]	42.00±3.79[b]	0.062
Burrow length (mm)	43.4±9.4	69.5±13.1	90.0±17.7	0.135
Burrow depth (mm)	38.0±8.1	60±10.1	76.7±13.6	0.114
Burrow diameter (mm)	12.3±0.3[b]	18.9±2.0[ab]	24.0±2.1[a]	0.007
Burrow volume (cm^3)	5.08±0.91[b]	19.77±5.66[ab]	43.97±13.91[a]	0.051
Highest number of burrow (nos.)	90	115	49	
Characteristics	Underwater during high tide but exposed to sunlight during low tide. Extensive salt tolerant grass present.	Shallow intertidal or higher intertidal ditch within mangrove, contains salt water during spring tide, mangrove roots and sea weeds present.	Higher intertidal, inundation during spring tide. Mangrove plants, eg. *Avicennia spp.* and *Acanthus spp.*, *roots of mangrove*	

Table 2. Bivariate analysis of Pearson correlation between sea to land ward distance and burrow characters (N=18 quadrates, 10 burrows from each quadrate)

Correlation		Distance from saltmarsh	Burrow number	Burrow length	Burrow Diameter
Distance from Saltmarsh	Pearson Correlation	1			
	Sig. (2-tailed)				
Burrow number	Pearson Correlation	-0.523			
	Sig. (2-tailed)	0.149			
Burrow length	Pearson Correlation	0.774[*]	-0.366		
	Sig. (2-tailed)	0.014	0.333		
Burrow Diameter	Pearson Correlation	0.811[**]	-0.119	0.782[*]	1
	Sig. (2-tailed)	0.008	0.761	0.013	

*Correlation is significant at the 0.05 level (2-tailed); ** Correlation is significant at the 0.01 level (2-tailed)

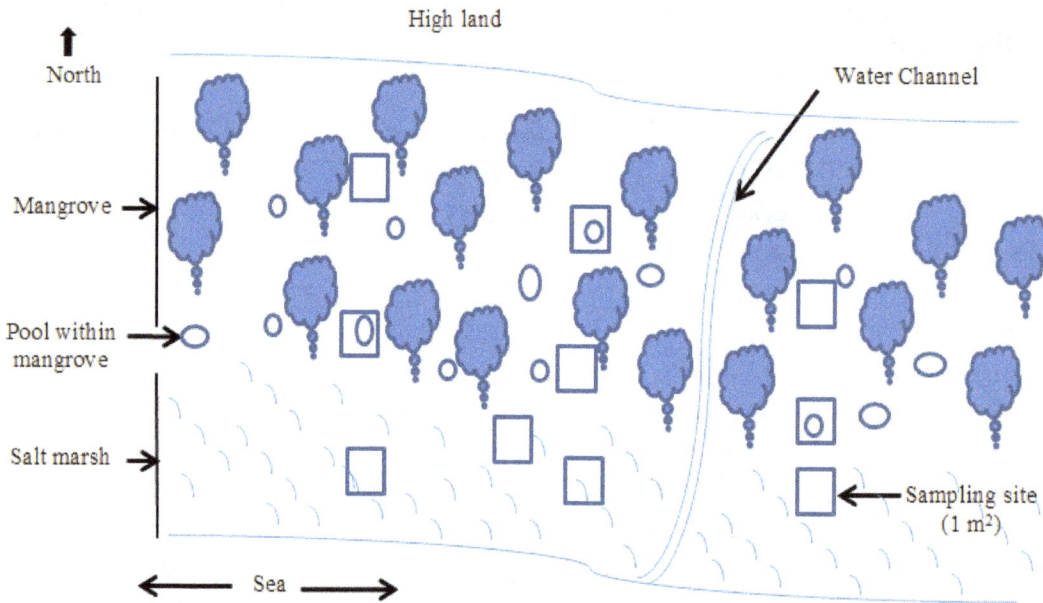

Figure 2. Schematic diagram of sampling sites in the saltmarsh, mangrove and mangrove pool. First three quadrates in saltmarsh were intentionally selected along the same line. Circle indicates pools intermingled with mangrove forest and squares (1m²) are randomly chosen quadrates

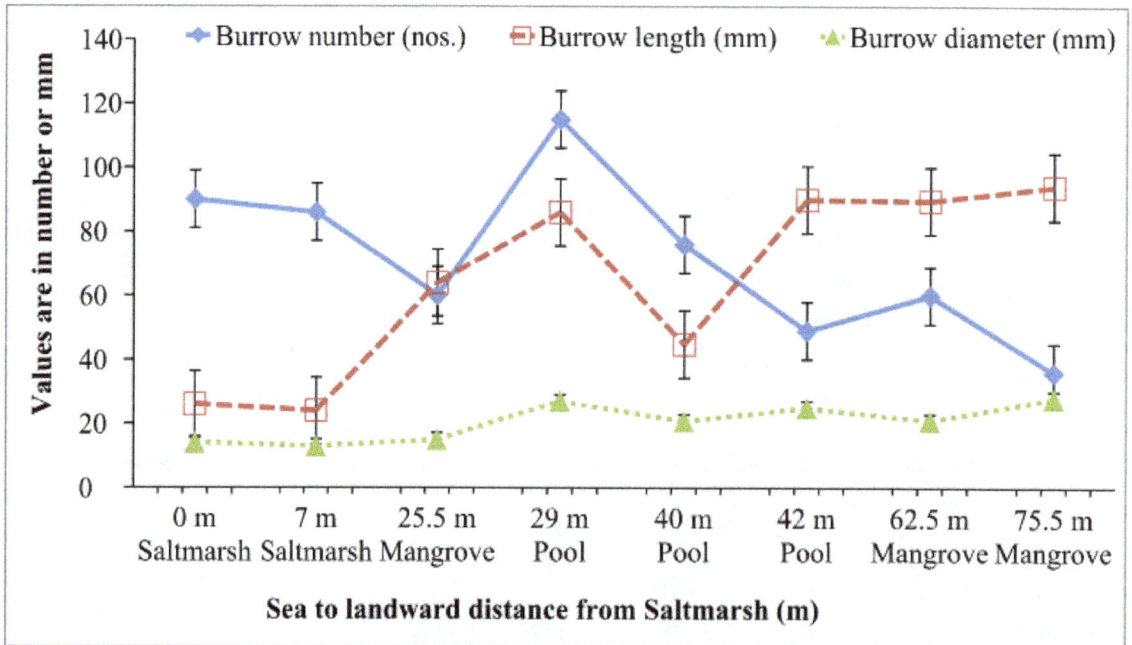

Figure 3. Comparison of Burrow characteristics along transect from saltmarsh (0 m) towards mangrove (every point indicates mean with error bar). Burrow density (blue) and burrow length (red) decreased with increasing distance, but burrow diameter (green) increased with increasing distance

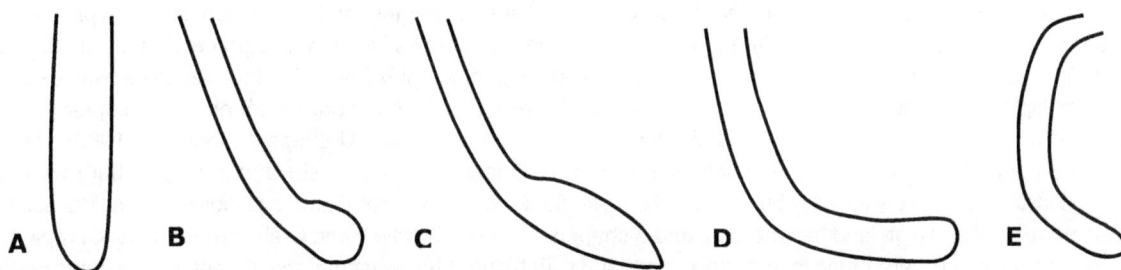

Figure 4. Burrow shapes: (A) I-shape, (B and C) J-shape, (D) L-shaped (E) C- shaped

DISCUSSION

Burrow abundance

Fiddler crab burrows were found in all the three sampling areas viz. saltmarsh, mangrove pool and mangrove highland with variable densities. Burrow distribution showed higher density (nos.) in mangrove pool and in saltmarsh than in mangrove high land (Table 1). Highest number of burrows (115 burrows m^{-2}) was found in mangrove pool (Table 1), where soil was relatively wet due to regular tidal splash and water containment, therefore, was soft for burrowing. Similarly, higher burrow density was also found in salt marsh which is relatively unstable area in terms of sedimentation due to regular tidal action. This area was usually occupied with early aged smaller fiddler crabs with higher density. In contrast, low burrow density was found in mangrove high land where soil was harder to make burrows. Mangrove high land is usually dried place inundated only in spring tide or highest high tide. So, fiddler crab distribution showed higher burrow prevalence in moist areas possibly due to the privilege of easy burrowing and wet condition required for gill moisture (Wolfrath, 1992; Lim and Diong, 2003). Soil characteristics (sediment) have important influence regarding burrow creation. Burrow density matches with the observation of Zeil and Hemmi (2006) who found fiddler crab live in large colonies that of all ages and sexes can reach densities of 50–100 individuals per square meter.

Burrow characters and morphology

Burrows in the saltmarsh are characterized by lower depth and smaller diameter. On the contrary, burrow depth and burrow length increased towards mangrove pool and mangrove high land (Figure 3). In mangrove high land, burrow length was higher (90 mm) with significantly higher burrow diameter, but burrow density was lowest among the studied habitats. Qureshi and Saher (2012) reported burrow length 105.4±45.9 mm and mean burrow diameter 13.7±3.0 mm for *U. annulipes*, which is similar with the present study. Therefore, with closeness to land, water availability decreased as this area is inundated only in spring tide and the soil become harder to burrow. In contrast, burrow depth increased with decreasing burrow density (Edmond *et al.,* 1996; Qureshi and Saher, 2012; Tina *et al.,* 2015). It reflects larger holes are relatively in highlands towards land (Lim and Diong, 2003) and perhaps burrows were created by larger individuals for breeding purpose (Hyatt and Salmon, 1979). Breeding burrows are relatively large and with hoods (Christy, 1982). Deeper burrows in high shores might be a way of maintaining lower burrow temperatures in areas that are exposed to sunlight for greater periods of time (Wolfrath, 1992; Lim and Diong, 2003). Reciprocally, in salt marsh (lower shore), burrow depth was lower along with significantly smaller burrow diameter which indicate smaller individuals were distributed close to the shore with higher densities. Therefore, these smaller burrows in anoxic sediments are significantly shorter in depths, which would help to improve aeration after every tidal flushing (Lim and Diong, 2003). Similar to mangrove land, mangrove pool was dominated with burrows with higher burrow length and diameter due to availability of water and easy burrowing. Therefore, water and sediment characteristics influence burrow density (Genoni, 1991).

Shape observation revealed *Uca* produce nearly vertical and straight and unbranched, mostly typical I-and J-shaped burrows (Tina *et al.*, 2015) and few L and C-shaped burrows with lower region enlarged (Montague, 1980). Similar burrow shapes were also observed in vertical and complex branching morphologies with a single and multiple entrances by Qureshi and Saher (2012). Different species of fiddler crab produce simpler L or J-shaped (Katz, 1980; Genoni, 1991, and Montague, 1980) burrow or even U-shaped (Montague 1980) and Y-shaped (Qureshi and Saher, 2012) complex burrows. Saltmarsh had the most I and J shaped burrows with simple structure possibly for easy burrowing by smaller fiddler crabs and short time availability within tidal action. Mangrove pool and high land had mostly I, and J-shaped burrows with few L and C-shaped burrows. Complexity of burrow structure were found in mangrove high land with higher burrow length and diameter represents larger fiddlers make deeper burrow in firm soil.

Burrow diversity along transect

Considering transect from salt marsh to mangrove land, burrow density decreased with increasing distance from shore toward mangrove (Edmond *et al.*, 1996). Reversely, burrow length and diameter increased landward, presumably larger individuals create stable burrows for breeding purpose in the higher intertidal zone and smaller individuals dominate in saltmarsh near water area. Moreover, burrows were deeper in drier upper shore land than in moist lower shore sediment (Takeda and Kurihara, 1987; Wolfrath, 1992; Lim and Diong, 2003). Burrow diameter increased with increasing distance and correlation analysis between burrow diameter and landward distance showed strong relationship (R=0.771, y = 0.1157x + 12.06). So, closer to saltmarsh, lower the burrow diameter and reciprocally near the land, higher the burrow diameter. Likewise, borrow length increased with increasing distance from shore showed positive correlation (Table-2). Thurman (1984) also showed that burrows of *U. subcylindrica* increase in depth with increasing distance from low water mark in Laguna Salado, Mexico. Therefore, shore to landward distance comparison explicit positive relation among distance, burrow length and burrow diameter. Contrarily, correlation between burrow number (density) and distance was negative. Therefore, higher number of burrows in saltmarsh was closer to sea water that decreased toward mangrove land. But, the higher numbers of burrows were abundant in mangrove pools. Pools are characterized by a small ditch contains salt water after being inundated with flood tide. Therefore, fiddler crab burrows were dominated in the margin of salt marsh and mangrove pool. So, water content (Frith and Brunenmeister, 1980; Ewa-Oboho, 1993) salinity and substratum (MacIntosh, 1989; Frusher *et al.*, 1994) of habitat have role in the distribution of fiddler crab burrows. Costa (2000) found positive correlations with water content and distribution of burrows for *U. thayeri* and Tina *et al.* (2015) observed burrows were shorter in length, shallower in depth, and smaller in volume at 2 m distance, but these measures increased with increasing distance from the river edge. In fine, it reflects complex interaction of multiple factors play role in burrow structuring, ie. soil texture sediment characteristics, water availability and burrowing individuals size (Thurman, 1987; Costa, 2000).

CONCLUSION

Fiddler crab burrow distribution and characters differed in different habitats of the intertidal coast of Bangladesh with more burrow density in shoreward and contrariwise towards landward. The number of burrows reduced towards higher intertidal zone but burrows of this area had higher burrow depth and diameter which represents water availability, substrate characteristics and size of burrowing individuals determine the burrow characters. Contrarily, smaller and temporary burrows were denser in saltmarsh towards shore. Highest burrows in mangrove pool indicated crabs prefer to stay in moist condition, where water availability support burrowing and gill activity. Fiddler crabs are ecologically important bio-turbator contribute in nutrient replenishment and support for nutrient enrichment in the intertidal habitat. The current study focused on fiddler crab burrows at the scale of 1 m^2 sampling plots. For better understanding of burrow characters, a more detailed study with a finer sampling design and massive burrow observation along with more intensive ecological study is required.

COMPETING INTEREST

The authors declare no conflict of interest.

ACKNOWLEDGMENTS

The authors would like to thank Dean, Faculty of Fisheries, Chittagong Veterinary and Animal Sciences University to support grant for this research and the students who participated in burrow observation in the intertidal coast.

REFERENCES

1. Aysha A, Kamal AHM, Mishra M, Nesarul MH, Padhi BK, Mishra SK, Islam MS, Idris MH and MB Masum, 2015. Sediment and carbon accumulation in sub-tropical salt marsh and mangrove habitats of north-eastern coast of Bay of Bengal, Indian Ocean. International Journal of Fisheries and Aquatic Studies, 2: 184-189.

2. Bertness MD and T Miller, 1984. The distribution and dynamics of Uca pugnax (Smith) burrows in a New England saltmarsh. Journal of Experimental Marine Biology and Ecology, 83: 211–237.

3. Bertness MD, 1985. Fiddler crab regulation of Spartina alterniflora production on a New England salt marsh. Ecology, 66: 1042–1055.

4. César II, Armendariz LC and RV Becerra, 2005. Bioecology of the fiddler crab Uca uruguayensis and the burrowing crab Chasmagnathus granulatus (Decapoda, Brachyura) in the Refugio de Vida Silvestre, Bahía Samborombón, Argentina. Hydrobiologia, 545: 237-248.

5. Chowdhury MSN, Hossain MS, Mitra A and P Barua, 2011. Environmental functions of the Teknaf Peninsula mangroves of Bangladesh to communicate the values of goods and services. Mesopatamian Journal of Marine Science, 26: 20-27.

6. Christy JH, 1982. Burrow structure and use in the sand fiddler crab, Uca pugilator (Bosc). Animal Behaviour, 30: 687–694.

7. Christy JH, Backwell PRY and S Goshima, 2001. The design and production of a sexual signal: Hoods and hood building by male fiddler crabs Uca musica. Behaviour, 138: 1065-1083.

8. Costa TM, 2000. Ecologia de caranguejos semiterrestres do gênero Uca (Crustacea, Decapoda, Ocypodiae) de uma área de manguezal, em Ubatuba (SP). PhD thesis, University Estadual Paulista.

9. Crane J, 1975. Fiddler Crabs of the World (Ocypodidae: Genus Uca). Princeton University Press, Princeton, New Jersey, xxiii 736 pp.

10. Edmond C, Mouton Jr. and DL Felder, 1996. Burrow Distributions and Population Estimates for the Fiddler Crabs Uca spinicarpa and Uca longisignalis in a Gulf of Mexico Salt Marsh. Estuaries, 19: 51-61.

11. Ewa-Oboho IO, 1993. Substratum preference of the tropical estuarine crabs, Uca tangeri Eydoux (Ocypodidae) and Ocypode cursor Linne (Ocypodidae). Hydrobiologia, 271: 119-127.

12. Frith, DW and S Brunenmeister, 1980. Ecological and population studies of fiddler crabs (Ocypodidae: genus Uca) on a mangrove shore at Phuket Island, western Peninsular Thailand. Crustaceana, 39: 157-184.

13. Frusher SD, Giddings RL and TJ Smith III, 1994. Distribution and abundance of Grapsid crabs (Grapsidae) in a mangrove estuary: effects of sediments characteristics, salinity, tolerance and osmoregulatory ability. Estuaries 17: 647-654.

14. Genoni GP, 1991. Increased burrowing by fiddler crabs Uca rapax (Smith) (Decapoda: Ocypodidae) in response to low food supply. Journal of Experimental Marine Biology and Ecology, 147: 267-285.

15. Herrnkind WF, 1968. The breeding of Uca pugilator (Bosc) and mass rearing of the larvae with comments on the behavior of the larval and early crab stages (Brachyura, Ocypodidae). Crustaceana, 2: 214-224.

16. Hodgson AN, 1987. Distribution and abundance of the macrobenthic fauna of the Kariega estuary. South African Journal of Zoology, 22: 153–162.

17. Hyatt GW and M Salmon, 1979. Comparative statistical and information analysis of combat in fiddler crabs, Uca pugilator and U. pugnax. Behaviour, 68: 1-23.

18. Icely JD and DA Jones, 1978. Factors affecting the distribution of the genus *Uca* (Crustacea: Ocypodidae) on an East African shore. Estuarine, Coastal and Shelf Science, 6: 315-325.

19. Katz LC, 1980. The effects of the burrowing of the fiddler crab *Uca pugnax* (Smith). Estuarine, Coastal and Shelf Science, 4: 233–237.

20. Lim SSL and CH Diong, 2003. Burrow-morphological characters of the fiddler crab, *Uca annulipes* (H. Milne Edwards, 1837) and ecological correlates in a lagoonal beach on Pulau Hantu, Singapore. Crustaceana, 76:1055-1069.

21. MacIntosh DJ, 1989. The ecology and physiology of decapods of mangrove swamps. Proceedings of the Zoological Society of London 59: 325-341.

22. Milner RNC, Booksmythe I, Jennions MD and PRY Backwell, 2010. The battle of the sexes? Territory acquisition and defence in male and female fiddler crabs. Animal Behaviour, 79: 735–738.

23. Mokhtari M, MA Ghaffar, G Usup and ZC Cob, 2016. Effects of fiddler crab burrows on sediment properties in the mangrove mudflats of Sungai Sepang, Malaysia. Biology, 5: 7.

24. Montague CL, 1980. A natural history of temperate western Atlantic fiddler crabs (genus *Uca*) with reference to their impact on the salt marsh. Contribution in Marine Science, 23: 25-55.

25. Nobbs M, 2003. Effects of vegetation differ among three species of fiddler crabs (*Uca* spp.). Journal of Experimental Marine Biology and Ecology, 284: 41-50.

26. Peer N, Miranda NAF and R Perissinotto, 2015. A review of fiddler crabs (genus Uca Leach, 1814) in South Africa. African Zoology, 50: 187-204.

27. Qureshi NA and NU Saher, 2012. Burrow morphology of three species of fiddler crab (*Uca*) along the coast of Pakistan. Belgium Journal of Zoology, 142: 114-126.

28. Rosenberg MS, 2001. The systematics and taxonomy of fiddler crabs: a phylogeny of the genus *Uca*. Journal of Crustacean Biology 21: 839–869.

29. Skov MW and RG Hartnoll, 2001. Comparative suitability of binocular observation, burrow counting and excavation for the quantification of the mangrove fiddler crab *Uca annulipes* (H. Milne Edwards). Hydrobiologia, 449: 201-212.

30. Takeda S and Y Kurihara, 1987. The distribution and abundance of *Helice tridens* De Haan (Crustacea, Brachyura) burrows and substratum conditions in a north eastern Japan salt marsh. Journal of Experimental Marine Biology and Ecology, 107: 9–19.

31. Thurman CL, 1984. Ecological notes on fiddler crabs of south Texas, with special reference to *Uca subcylindrica*. Journal of Crustacean Biology, 4: 665-681.

32. Thurman CL, 1987. Fiddler crabs (genus Uca) of eastern Mexico (Decapoda, Brachyura, Ocypodidae). Crustaceana, 53: 94-105.

33. Tina FM, Jaroensutasinee M, Sutthakiet O and K Jaroensutasinee, 2015. The fiddler crab, *uca bengali crane*, 1975: population biology and burrow characteristics on a riverbank in southern Thailand. Crustaceana, 88: 791-807.

34. Warren JH, 1990. The use of open burrows to estimate abundances of intertidal estuarine crabs. Australian Journal of Ecology, 15: 277-280.

35. Wolfrath B, 1992. Burrowing of the fiddler crab *Uca tangeri* in the Ria Formosa in Portugal and its influence on sediment structure. Marine Ecology Progress Series, 85: 237–243.

36. Zeil J and JM Hemmi, 2006. The visual ecology of fiddler crabs. Journal of Comparative Physiology, 192: 1–25.

17

PRESENT STATUS AND PROBLEMS OF FISH SEED MARKETING IN SYLHET DISTRICT, BANGLADESH

Shahrear Hemal[1], Md. Shahab Uddin[1], Md. Saif Uddin[2], Bhaskar Chandra Majumdar[3*], Md. Golam Rasul[3] and Md. Tariqul Alam[1]

[1]Department of Aquaculture, Sylhet Agricultural University, Sylhet-3100, Bangladesh; [2]Department of Aquaculture, Bangabandhu Sheikh Mujibur Rahman Agricultural University, Gazipur-1706, Bangladesh; [3]Department of Fisheries Technology, Bangabandhu Sheikh Mujibur Rahman Agricultural University, Gazipur-1706, Bangladesh

*Corresponding author: Bhaskar Chandra Majumdar; E-mail: bhaskar.bsmrau@gmail.com

---

ARTICLE INFO

Key words

Fish seed
Marketing
Stakeholders
Income
Transportation
Mortality

ABSTRACT

This study was conducted to explore the present status and problems of fish seed marketing system in Sylhet district, Bangladesh. Data were collected through questionnaire interview from the selected areas during April to September 2016. Brood fishes were collected from wild sources as well as hatchery produced brood fishes also used for seed production. Good length, weight and age of brood fishes were selected for spawning and induced breeding. In nursery, hatchlings were reared for 30-40 days and 37.7% nursery owner practiced single cycle production/year where the average stocking density of seed was found 24.65±3.94 g/decimal (mean±SD). Six different fish seed marketing channels were identified where hatchery owners, nursery owners, forias (retailer) and fish farmers were main stakeholders. The highest (6520 Tk/day) and lowest (355 Tk/day) average income were found in hatchery owner and fish farmer, respectively. Oxygenated bag, big aluminum bowl/container and plastic barrels with continuously agitate the water were used for seed transportation. Maximum 17.67% seed mortality was noticed in hatchery owner and minimum 5.67% in fish farmers. Late breeding season, lack of capital, lack of technical knowledge on hatchery operation and management, poor transport facilities, high transportation cost, high labor cost, lack of training and high price of spawn are identified as some major problems.

INTRODUCTION

Bangladesh is a riverine country having lots of rivers, canals, floodplains, ponds, beels, haors, reservoir, manmade lakes and a long coastline etc. These waterbodies are valuable sources of fish. Both capture and culture fisheries play an important role for animal protein supply in Bangladesh. Aquaculture sector of Bangladesh is expanded rapidly with respect to both the quantity and variety of species. In the year of 2014-2015, total fish production of the country was 3.68 million MT of which inland aquaculture sector provides 2.06 million MT (DoF, 2016). Development of aquaculture is influenced by good quality seed and good quality feed. In the year of 1985, spawn production from natural and hatchery sources were 19,362 kg and 4,962 kg (Hussain and Mazid, 2001). The demand of substantial supply of quality fish fry and fingerlings are increased day by day. At present private fish hatcheries and Government Fish Seed Multiplication Farm (FSMFs) with hatchery facilities contributed 5,36,983 kg and 10,566 kg hatchlings respectively (FRSS, 2016). Department of fisheries (DoF) of Bangladesh is encouraging people to enhance the fish seed production by setting up hatcheries and nurseries. On the other hand 800 private hatcheries all over the country mostly help fish farmer to reduce the dependency on nature for fish seed. This large numbers of hatcheries created a new marketing system all over the country to reach their fish seed to fish farmer.

A market system is the network of buyers, sellers and other actors that come together to trade in a given product or service. In this marketing system there are different actors such as nursery owners, foria, hawkers, wholesalers, fry traders etc. The marketing channels are the alternative courses of item flows from producers to consumers (Kohls and Uls, 1980). Marketing channel may be small or long for a particular product depending on the type and quality of the relevant factors. Fish seed marketing of the country largely depends on private sector and a large number of people are associated with this system for their livelihoods. But, stress from transportation caused 6200 MT fish seed mortality in a year over the country (Hasan and Brat, 2006). Besides, lack of capital, lack of technical knowledge, high lease value, high price of production inputs, violence, intense market competition, the lack of policy support etc. have impact on stakeholders on this marketing system. Sylhet (north western district of Bangladesh) district is endowed with numerous natural inland water bodies such as rivers, canals, ponds, beels etc. A special wetland ecosystem of the country is found in sylhet region named haor having huge possibility of aquaculture. There are about 5 fish hatcheries and 152 fish nurseries in the Sylhet region. In 2014, two government fish seed multiplication farm, one government nursery, private hatcheries and nurseries are produces 950 kg hatchling and 81.94 million fry and fingerling (DoF, 2005). But, the marketing channel and problems associated with marketing of fish seed in Sylhet region has not been yet quantified. The objectives of the study were to know about the existing fish seed marketing system and identify the problems of the fish seed marketing system in Sylhet region.

MATERIALS AND METHODS

Selection of study area, Target groups and Period of the study

Four upazillas of Sylhet district were selected for this study and the target groups were hatchery and nursery owners, fish seed hawkers as well as fish farmers. Production season of fish seed generally started from 1st week of April and ends up by 1st week of September every year. However, the survey was carried out for six months from April to September 2016.

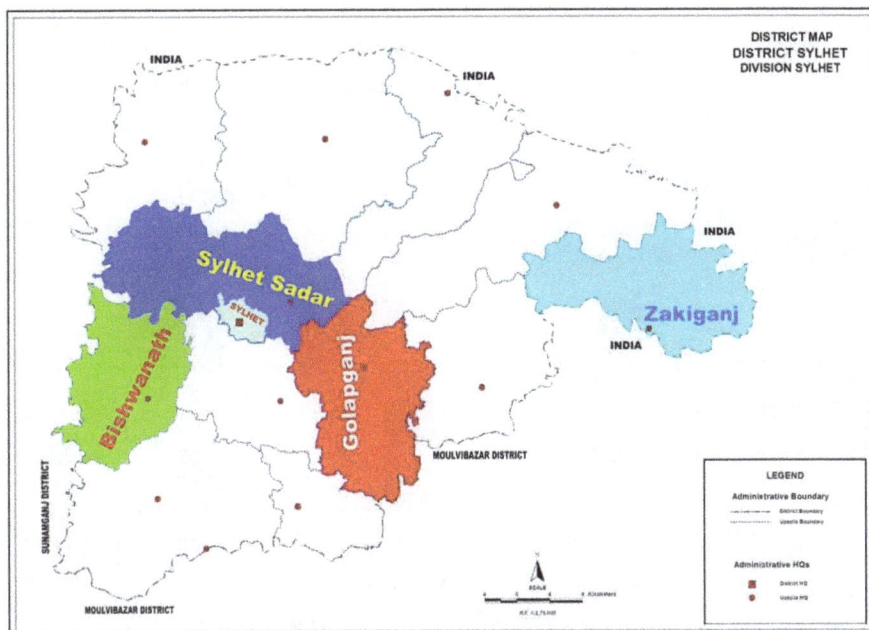

Figure 1. A map of selected study areas of Sylhet District. (*source: www.lged.gov.bd*)

Selection of sample and sampling techniques

In order to meet the objectives of the study, different types of fish seed and fingerling producers and hawker were selected. The sample size for the present study is given in Table 1.

Table 1. Sample size of the present study

Types of stakeholders	Sylhet District				Total
	Sylhet Sadar	Golapganj	Zakiganj	Bishwanath	
Hatchery owner	1	2	2	2	7
Nursery owner	10	10	10	10	40
Fish seed hawker	10	10	10	10	40
Fish Farmer	10	10	10	10	40
Total					127

Collection of data, Data processing and analysis

Data were collected from the respondents by using the pre-prepared questionnaire interview. All the collected data were summarized and scrutinized carefully and recorded. Data were presented mostly in the tabular form, because it is simple in calculation, widely used and easy to understand.

RESULTS AND DISCUSSION

Hatchery Status

At present, there are only 5 private fish hatcheries and 2 Governments Fish Seed Multiplication Farm (FSMFs) in Sylhet district. The hatcheries are situated at Sylhet Sadar, Golapganj, Bishwanath, Osmaninagar and Zakiganj. Among 5 private hatcheries, one was Catfish hatchery, three was Carp hatchery and other one was Tilapia hatchery. During the study period spawn production was 431 kg carp fishes, 220 kg catfishes and 567 kg Tilapia (Figure 2). In the present study it was found that the carp hatchery activities start from 1st week of April and continued up to 1st week September. Peak seasons of seed production were late-April to mid-

June. It was also found that the Catfish hatchery activities start from 1st week of April and continued up to 1st week September. Peak seasons of seed production were late-April to mid-June. The only one Tilapia hatchery in Golapganj produced only tilapia fry throughout the year. After 4 days hatchling was fed feed treated with 17α -Methyltestosterone. The brood fishes were produced in the on hatchery.

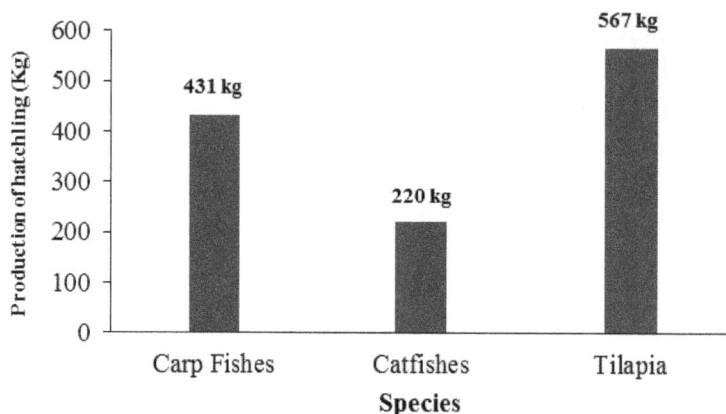

Figure 2. Hatchling production in Sylhet District

Above result was relevant with the Sharif and Asif (2015) reported that hatchlings production of Indian major carps were 24,720 kg, Exotic carps were 21754 kg and other species were 2966 kg respectively in 2013 at Jessore sadar. Rahaman (2007) also got that 85 hatcheries in Jessore area were produced 4211.5 kg hatchling at study period of which 80% were reared in the same area.

Facilities used in Hatchery

Smooth operation of hatchery mainly dependent on facilities of hatchery. Good production of hatcheries depends on appropriate number of brood ponds, breeding tanks with other facilities, cisterns, incubators, fry rearing tanks etc. Besides, there is difference between capabilities used in Carps Hatchery, Catfish Hatchery and Tilapia Hatchery. A short list of facilities used in hatcheries was shown in the Table 2.

Table 2. Lists of facilities used in hatchery

Facilities	Fish Hatchery	Tilapia Hatchery	Catfishes Hatchery
Brood fish pond	5-1	5	3
Breeding Tank	7-1	1	5
Cistern	8-1	-	-
Incuba Bottle	13-8	-	-
tor Tray	-	60	-
Hormone	Ovaprim, PG, HCG	17α-Methyltestosterone	Ovaprim, Ovatide
Brood Transport materials	Plastic drum, Aluminum patil	Aluminum patil	Aluminum patil

Brood management

A good number of spawn productions from hatchery mainly depends on the selection and successful management of brood fishes. Best quality and higher number of spawns can be obtained from the good quality brood fishes. The criteria and management activities for brood fishes were followed by the hatchery owner show in the Table 3.

Table 3. Criteria and management of brood fishes are maintained by the hatchery owner

Criteria		Indian Major Carps (Rui, Mrigal, Kalibaus, Gonia)	Catla	Exotic Carp	Tilapia	Catfishes (Shing & Magur)
Av. Wt.	Male	500 gm - 2kg	1200 gm to 3 kg	500 gm to 2 kg	400-50 kg	160 gm
	Female	700 gm - 2 kg	1400 gm to 3 kg	700 gm to 2 kg	250-500 gm	200 gm
Age	Male	2-3 years	3-4 years	2-3 years	2 years	3 years
	Female	2-3 years	3-4 years	2-3 years	2 years	3 years
Source of Brood		River, Own farm, Other farm		Own or Other farm	Own farm	Own farm
Feed for brood		24 hour composted mustard oil cake with rice bran and wheat bran or commercial feed				Commercial feed
Feeding rate		1-2 times in a day				3 times/night

This result supported the result of Dwivedi and Zaidi (1983) who reported that brood stock is a prerequisite for all types of hatchery production and proper brood stock management will lead to better breeding responses and increased fecundity, fertilization, hatching and larval survival rates and more viable fish seed. Sultan (2008) also identified that maximum hatchery owners (80%) in Chachra collected broods from the Halda and the Jamuna rivers to avoid inbreeding and all of the hatchery owners practiced induced breeding in their hatcheries.

Nursery operation

About 61.18% (93 of total 152 nurseries) nursery of Sylhet district are situated in study areas. Nursery of Sylhet district mainly produces fry or fingerlings of Rui, Catla, Mrigal, Bata, Kalbasu, Sarpunti, Bighead carp, Silver carp, Grass carp, Tilapia, Pangas etc. The size of the nursery depends on capital for nursery operation, land, knowledge about nursery operation etc. The highest 42.5% of nursery was found in a range of 0.5- 1.0 acre size while lowest 17.5% nursery was found in a range of 1.5-above acre (Figure 3).

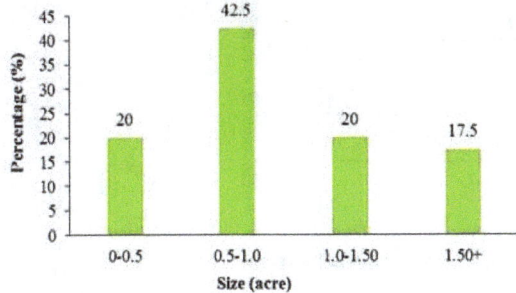

Figure 3. Size of nursery in acre

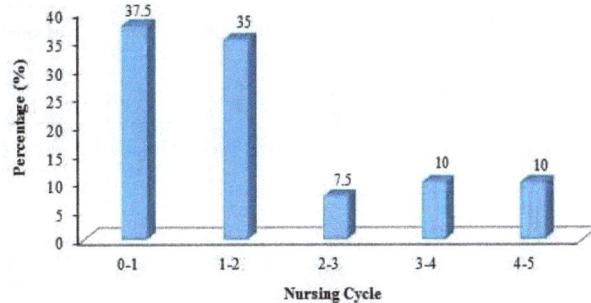

Figure 4. Number of nursery cycle by nursery owner per year

In this region nursing cycle varies from highest 4-5 cycle/year due to collection of spawn from outside of the district, high price of spawn than other region and nursing mostly for grow out of fish. It was observed that 37.7% farmers completed single cycle/year, 35% farmers completed 1-2 cycle/year and only 10% farmer completed 4-5 cycle/year (Figure 4).

Fish Farm

Fish hatchling, fry or fingerling are reared up to table size for selling in the market or home consumption. In the fish farm of Sylhet region mostly cultured species are Tilapia, Koi and carps like Rui, Catla, Kalbasu, Mrigal etc. The size of the farms depends on the economic condition of the farmer. The highest number of farms was covered area 2-4 acre (Figure 5).

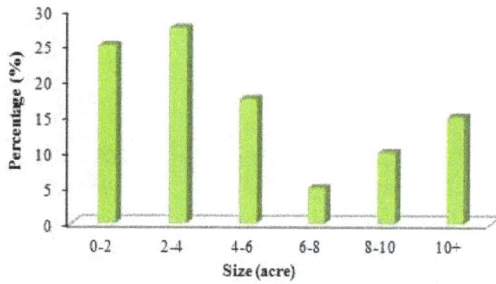

Figure 5. Size of fish farm in acre

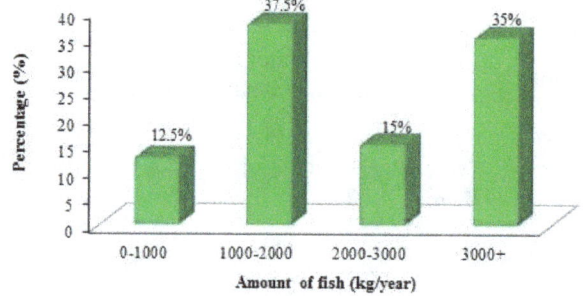

Figure 6. Total fish production in fish farms

Islam (2010) revealed that average pond size was 0.11 ha in Moulvibazar district. The most of the farmer (55%) found to operate nursery in their farm for fry or fingerlings to grow-out. The productions of farms relay on quality of fish seeds, management activities followed by the farmers, the size of the farms and knowledge of farmer on fish culture. The highest number of farm 37.5% produced fish between 1000-2000 kg/year (Figure 6).

Fish seed marketing channel

The marketing channels are the alternative routes of product flows from producers to consumers (Kohls and Uhl, 1980). Fish seed marketing channel starts from hatchery owners, passes through some intermediaries and ends with the fish farmers who culture table fish. In the present study, fish seed marketing channels lunched with the brood fish collection to fish farmers through proper channels (Figure 7).

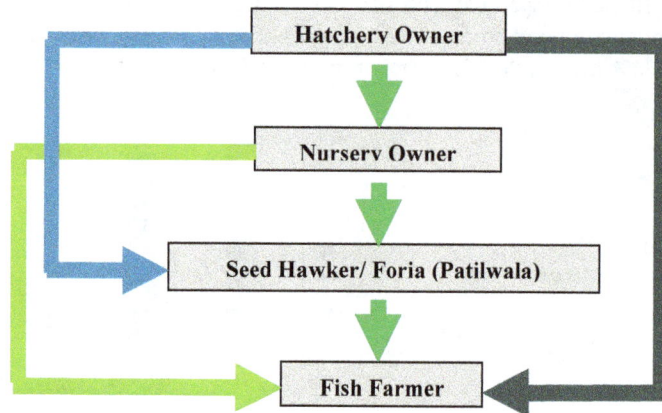

Figure 7. Flow chart of fish seed marketing channels in Sylhet district

Channels of commercial fish seed marketing in Sylhet area

About Six different fish seed marketing channels were identified in Sylhet area. Among them, channel-3 was best. It's because, there is no middleman between hatchery owners and fish farmers. That's why; fish farmers can buy their quality seeds in low price. Abdulla-Al-Asif et al., (2015) reported that marketing channel of fish fry and fingerling is start with brood pond and continues with hatchery, nursery, fry and fingerling traders, intermediates, buyer, farmer, then farming pond or rearing pond. Malek (2007) identified seven marketing channels where hatchery and nursery owners, wholesalers, forias and fish farmers were major stakeholders. Similar findings were reported by Sharif and Asif (2015).

Channel-1:	Hatchery owners	⟶ Nursery owners ⟶ Forias ⟶ Fish Farmer
Channel-2:	Hatchery owners	⟶ Nursery owners ⟶ Fish Farmer
Channel-3:	Hatchery owners	⟶ Fish Farmer
Channel-4:	Nursery owners	⟶ Forias ⟶ Fish Farmer
Channel-5:	Nursery owners	⟶ Fish Farmer
Channel-6:	Forias	⟶ Fish Farmer

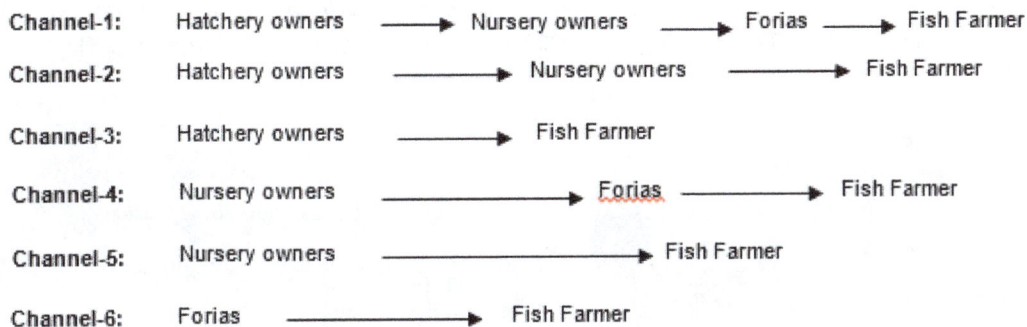

Transportation

Fish seed or fingerlings were found to transport by vehicle like pick-up, auto-rickshaw, bus, truck, van etc. About 80% foria (retailer) used bus to transport fry or fingerlings to the district and sell the fingerlings to the fish farmers door to door on foot hanging the aluminums bowl/container on their shoulder. Pickups and auto-rickshaws were found to be the main means of transportation of fry to the distance places. Fish spawns were transported in polythene bag by using auto-rickshaw, pickup and truck. Brood fish were transported by aluminum patil and plastic drum by pickup or vans. Similar result was showed by Haque *et al.* (1991), Abdulla-Al-Asif *et al.* (2015) and Sultan (2008).

Figure 8. Mortality of fish seed of different stakeholders in marketing channel

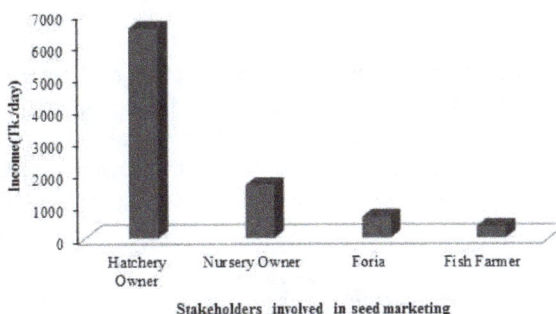

Figure 9. Average income (Tk/day) of Stakeholders involved in fish seed marketing

Mortality of fry due to transportation

It was observed that fries were transported by polythene bag and aluminum patils. Polythene bag were filled with oxygenated. Fries in aluminum patils were transported with agitation by hand and change water. Sometimes due to high density, longer duration and improper transportation method caused mass mortality which result into loss of overall business. In the study area average mortality was 17.67%, 13.28%, 9.5% and 5.67% for hatchery owners, nursery owners, forias and fish farmer respectively (Figure 8). Malek (2007) found average mortality was 15.67%, 11.25%, 8.5%, and 4.67% for hatchery owners, nursery owners, wholesalers and forias respectively in the seed marketing channel in Mymensingh area.

Income of different stakeholders in fish seed marketing channel

The income of hatchery and nursery owners in fish seed marketing system is more than the income of the fry traders, wholesalers and fish farms. The average income of hatchery owners were found at Tk.6,520 day/hatchery owner, average net profit of nursery owners were estimated at Tk. 1,646/day, average income of forias were estimated Tk. 655/day and average profit of fish farmers were estimated Tk. 355/day (Figure 9).

Malek (2007) found similar result that the average income of hatchery owners were Tk.8,375 day/hatchery owner, nursery owners were Tk. 1,660/day, Tk. 1,362.17/day for wholesalers and forias were Tk. 490/day in Mymensingh area.

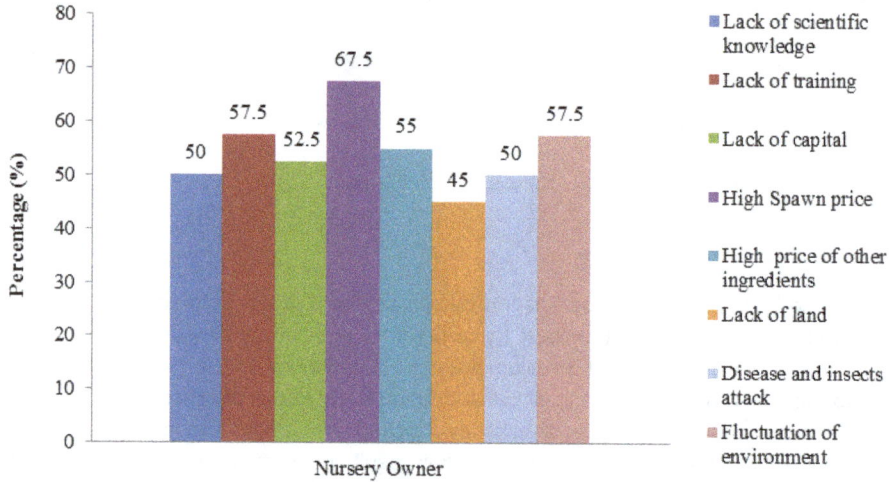

Figure 10. Problems faced by nursery owners

Problems and constraints of production and marketing of fish seed

Major problems of nursery owners includes- high price of spawn, lack of training on nursery operation, fluctuation of environmental condition, diseases of fingerlings, price fluctuation etc. Among the problems, high price of spawn was highest (67.5%), followed by lack of training (57.5%) and fluctuation of environment (57.5%) were crucial constrains for nursery operation (Figure 10)

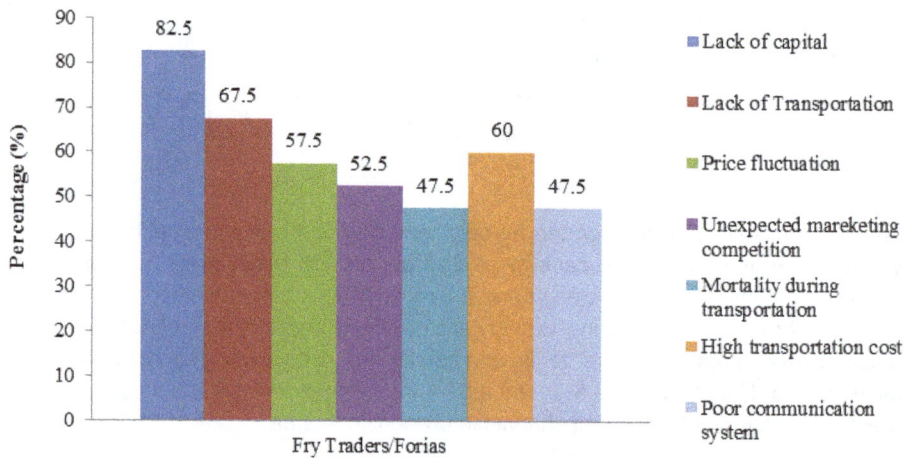

Figure 11. Problems faced by fry traders/retailers

Fry retailers maintained link between the producer (Hatchery owner, Nursery owner) and Fish farmer. The problems which are faced by the fry retailers/forias include lack of capital (82.5%) was most important. Among others lack of transport facilities (67.5%) and high transportation cost (60%) could be mentioned (Figure 11).

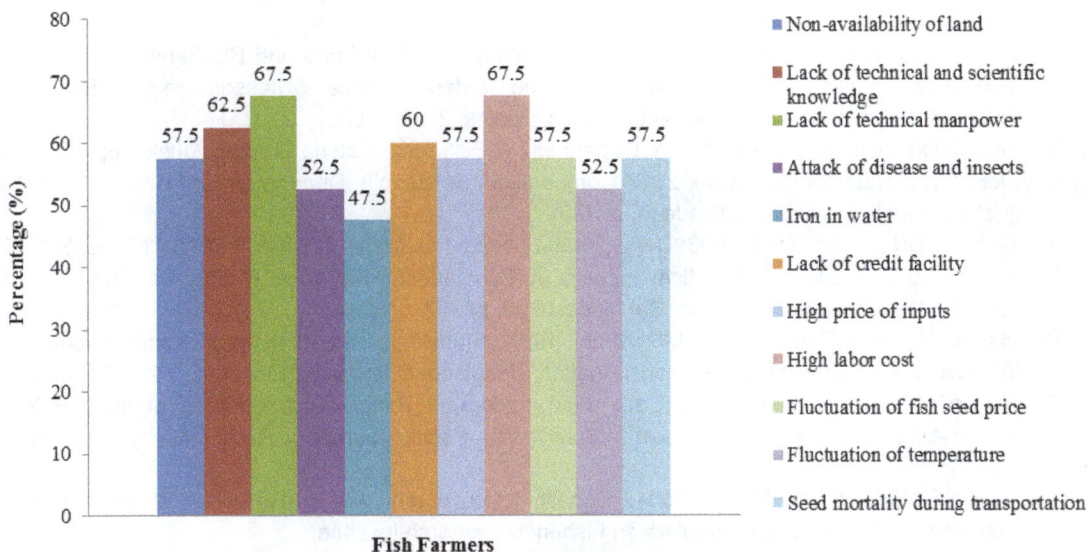

Figure 12. Problems faced by fish farmers

In this region 55% fish farmer collected seed directly from hatchery. Most of the farmer collected seed from outside of the district. In our study, it was observed that fish farmers faced a lots of problems. Among them, lack of technical man-powers (67.5%) and high labor cost (67.5%) were noticeable problems (Figure 12).

Similarity was found by Malek (2007) stated that problems faced by hatchery owners in Mymensingh area included lack of disease tolerant brood, inbreeding problems, lack of technical knowledge, price fluctuation and unexpected market competition. On the other hand, the lack of capital, lack of transportation facilities was the main problems for nursery owners. Rahaman et al., (2007) also stated that the main constrains were lack of capital (35%) followed by lack of technical knows-how (27%) faced by fry traders. The nursery operators faced the problems like lack of capital (31%), high lease value (25%), high price of production inputs (17%), violence (12%), intense market competition (9%) and the lack of policy support (6%) in Jessore. Sharif and Asif (2015) also claimed *Argulus* disease as the main problem, 95% of hatchlings mortality is caused by Argulus disease in Jessore.

CONCLUSION AND RECOMMENDATION

Aquaculture activities are rapidly increased in the Sylhet region. The supply of good quality fry and fingerlings for successful aquaculture activities depends on a good seed marketing system. But, some problems are evident in the marketing channel and transportation system of fry. Traditional transportation system, involvement of multiple middlemen, high fry mortality, lack of technical knowledge of hatchery and nursery operators are major problems. For developing an efficient fish seed marketing the following recommendations should be followed through participatory approaches with hatchery owners, farmers, traders, government agencies, and NGO stakeholders. In this regard, quality brood should provide to the hatchery owners by government arrangement through establishment of live brood banks. Formal fry and fingerling trading networks should be developed locally and regionally by the government and other developing partners so that fry and fingerling producers and farmers can get their actual benefit. To produce quality brood stock, quality seed and an effective seed marketing system, institute-industry research partnership should establish along with hatchery and nursery operators.

REFERENCES

1. Abdulla-Al-Asif M, A Samad, MH Rahman, M Almamun, SMY Farid and BS Rahman, 2015. Socio-economic condition of fish fry and fingerling traders in greater Jessore region, Bangladesh. International Journal of Fisheries and Aquatic Studies, 2: 290-293.

2. DoF, 2005. Matsha pakka sankalon. Department of Fisheries, Matsha Bhaban, Dhaka. pp. 133-139.

3. DoF, 2016. National Fish Week 2016 Compendium (In Bengli). Department of Fisheries, Ministry of Fisheries and Livestock, Bangladesh. p. 148.

4. Dwivedi, SN and GS Zaidi, 1983. Development of carp hatcheries in India. Fishing Chimes, 3: 1-19.

5. FRSS, 2016. Fisheries Statistical Report of Bangladesh. Fisheries Resources Survey System (FRSS), Department of Fisheries, Bangladesh. 32: pp. 57.

6. Haque MZ, MA Rahman and MS Shah, 1991. Studies on the density of Rohu (*Labeo rohita*) fingerlings in polythene bags for transportation. Bangladesh Journal of Fisheries, 14: 145-148.

7. Hasan M and AN Bart, 2006. Carp seed traders in Bangladesh: Sources of livelihoods and vulnerability resulting from fish seed mortality. Asia-Pacific Journal of Rural Development, 16: 101-124.

8. Hussain MG and MA Mazid, 2001. Genetic improvement and conservation of carp species in Bangladesh. Mymensingh: Bangladesh Fisheries Research Institute.

9. Islam S, 2010. Studies on pond fish farming and livelihoods of rural farmers in some selected areas of Maulvibazar district. M. S. Thesis. Department of Aquaculture, Bangladesh Agricultural University, Mymensingh, pp. 36.

10. Kohls RL and JN Uhl, 1980. Markerting of Agricultural products (5th edition), Macmillan publishing Co. Inc., New York.

11. Malek MA, 2007. Fish seed marketing in Mymensingh area, MS Thesis, Department of Aquaculture, Bangladesh Agricultural University, Mymensingh, pp. 64.

12. Rahaman MM, MA Sayeed, A Paul and M Nahiduzzaman, 2007. Problems and Prospects of Fish Fry Trade in Jessore District of Bangladesh. Progressive Agriculture, 18: 199-207.

13. Sharif BN and AA Asif, 2015. Present status of fish hatchlings and fry production management in greater Jessore, Bangladesh. International Journal of Fisheries and Aquatic Studies, 2: 123-127.

14. Sultan MS, 2008. Present status of hatchery operation and fry marketing system at Chachra, Jessore. MS Thesis, Department of Aquaculture, Bangladesh Agricultural University, Mymensingh, pp. 54.

18

SPECIES AVAILABILITY, CULTURE TECHNIQUE, REPRODUCTION OF PRAWN AND SHRIMP IN BANGLADESH

Sharmin Akter Shampa[*], Nusrat Nasrin, Marufa Khatun and Salma Akter

Department of Fisheries Biology and Genetics, Faculty of Fisheries, Bangladesh Agricultural University, Mymensingh-2202, Bangladesh

*Corresponding author: Sharmin Akter Shampa, E-mail: sharminshampa10@gmail.com

ARTICLE INFO	ABSTRACT
Key words Prawn Shrimp Species Culture Breeding	Bangladesh is considered one of the most suitable countries in the world for freshwater prawn farming because of its favorable agro climatic condition. The expansion of shrimp farming is triggered by the increased international market demand, seed production and intensive shrimp culture. Shrimp farming is having a positive impact on the livelihoods of many people in Bangladesh, especially the poorer farmers. Prawn and shrimp farming offer a reliable source of revenue that is often more profitable than other kinds of farming, or other non-farming employment. The objective of this paper was to review species availability, culture technique and artificial breeding of prawn and shrimp in Bangladesh.

INTRODUCTION

The freshwater ecosystems of Bangladesh provide a unique environment for enormous prawn production potential because of the favorable climate and availability of wild seed stock. Moreover, the populace of this country has close ties with the wetland systems including rivers, deltas, rice paddies and fish ponds making them naturally prepared to exploit the full potential of the freshwater prawn fisheries. The latest estimate of the area of land under shrimp cultivation has jumped from 64,000 *ha* in 1983 to 275,000 *ha* in 2012 (FRSS, 2013) in Bangladesh. In the coastal area of the greater Khulna region having a tropical climate, productive and unpolluted estuarine areas is considered to be a suitable natural habitat for penaeid shrimp culture. The culture of prawn and shrimp in Bangladesh has been drawing greater attention by fish farmers, particularly in brackish waters.

RESEARCH METHODOLOGY

This review collected information from different research articles and from places where prawn and shrimp cultures are abundant for example Khulna, Bagherhat, Jessore, Paikgacha, Koyra, Batiaghata, Dumuria, Rupsha, Terokhada, Digholia, and Fultala.

RESULTS

Species availability

There are 24 species of freshwater prawns including 10 species of *Macrobrachium* in Bangladesh. Among these species, only *Macrobrachium rosenbergii* is commercially cultured. There is a high prospect of *M. malcolmsonii* culture in Bangladesh since larval production and culture is similar to that of *M. rosenbergii*, which has been farmed successfully in the India. The available *Macrobrachium* and *penaeus* species in Bangladesh are shown in the following table:

Table 1. The available *Macrobrachium* species in Bangladesh (Ahmed et al., 2008; Ahamed et al., 2014)

Scientific name	English name	Local name
Macrobrachium birmanicus	Freshwater prawn	Thenguaicha
Macrobrachium dayanus	Freshwater prawn	Kairaicha
Macrobrachium dolichodactylus	Freshwater prawn	Icha
Macrobrachium lamarrei	Freshwater prawn	Icha
Macrobrachium malcolmsonii	Monsoon river prawn	Chotkaicha
Macrobrachium mirabilis	Freshwater prawn	Lutiaicha
Macrobrachium nipponense	Oriental river prawn	Icha/chingri
Macrobrachium rosenbergii	Giant freshwater prawn	Golda chingri
Macrobrachium rude	Freshwater prawn	Godaicha
Macrobrachium villosimanus	Freshwater prawn	Dimuaicha
Penaeus monodon	Giant tiger shrimp	Bagda chingri
Penaeus semisulcatus	Indian white shrimp	Sada icha
Penaeus indicus	Indian white shrimp	Sada icha
Penaeus japonicas	Kuruma prawn	Dora kata / Japani chingri
Penaeus merguiensis	Banana shrimp	Bagda chama
Penaeus penicillatus	Red tail prawn	Chama icha
Penaeus orientalis	White prawn	Boro chama

DISTRIBUTION

The giant tiger prawn is broadly distributed throughout the greater part of the Indo-Pacific region, extending from the northward to Japan and Taiwan, eastward to Tahiti, southward to Australia, and westward to Africa (Motoh, 1985). The species *Penaeus monodon* occurs mainly in Southeast Asian waters and it is widely distributed from 30°E to 155°E longitude and 35°N to 35°S latitude. In broadly, the *Penaeus monodon* is a widely distributed penaeid shrimp species which is native to the Indo-West Pacific with a range comprising southern Japan, China, Taiwan, Vietnam, Korea, the Philippines, Cambodia, Malaysia, Singapore, Australia, Thailand, Indonesia, Papua New Guinea, Myanmar, Bangladesh, Pakistan, India, Sri Lanka, Tanzania, Madagascar, and South Africa and the Red Sea off Yemen (FAO, 2012). Their main fishing grounds are located in the tropical states, predominantly in Malaysia, Indonesia and the Philippines. The peneideans are mainly abused in tropical, subtropical, and warm temperate waters, and distributed at all depths and in almost all environmental conditions. The fry, juveniles and adolescents inhabit surface waters such as in shallow water and mangrove estuaries, while most of the adults inhabit deeper waters down to almost 160 m.

Current status of prawn and shrimp farming

Production of shrimp and prawn was estimated from the Khulna district of Bangladesh, from the shrimp depots of Paikgacha, Dacope, Koyra, Batiaghata, Dumuria, Rupsha, Terokhada, Digholia, Fultala, and Metro Upazila in Khulna district. The total harvest of shrimp and prawn and observed production in shrimp depots in Khulna district was 21611 and 18620 ton respectively where the total area of shrimp farming is 58472 hector. The annual total harvest of shrimp was 5873, 4388, 1636, 2145, 5416, 616, 582, 569, 345 and 41 ton at Paikgacha, Dacope, Koyra, Batiaghata, Dumuria, Rupsha, Terokhada, Digholia, Fultala, and Metro Upazila in Khulna district respectively. The annual total shrimp production in depots was 4833, 3675, 1408, 1992, 4694, 602, 529, 537, 311 and 39 ton at Paikgacha, Dacope, Koyra, Batiaghata, Dumuria, Rupsha, Terokhada, Digholia, Fultala, and Metro Upazila in Khulna district respectively. Where the total shrimp farming area are 17276, 12680, 4530, 6253, 13284, 1178, 1102, 1070, 987 and 112 hector at Paikgacha, Dacope, Koyra, Batiaghata, Dumuria, Rupsha, Terokhada, Digholia, Fultala, and Metro Upazila in Khulna district respectively.

Production and its contribution to the economy of Bangladesh

The firstly records of Bangladesh's export of freshwater prawns dates back to the 1960's and was mainly from the capture fishery to markets in the USA, UK, France, Italy and Belgium (Ahamed et al., 2014). In the early 1990s, prawn farming had developed well especially in the southwestern Bangladesh; over 90% of the prawn exports came from the capture fisheries to confirm the importance of this subsector in the freshwater prawn fisheries of Bangladesh. The export of prawn and shrimp and foreign earning during 1999/2000 to 2011/2012 are presented in Figure 1.

Reproduction

Wild males produce spermatozoa from around 35 g BW and females becomes gravid from 70 g. Mating (Figure 2) occurs at night, shortly after moulting, while the cuticle is still soft, and sperm are subsequently kept in a spermatophore (sac) inserted inside the closed thelycum of the female. Females of *P. monodon* are highly fecund, Spawning occurs at night and fertilization is external, with females releasing sperm from the thelycum as eggs are released in offshore waters. Nauplii hatch 12–15 h after fertilization. Females spawn 4 times during their lifespan at carapace length of 50,62,66, and 72 mm. but it is unknown how many times males mate. This species breeds year-round. The range of the number of offspring is 248,000 to 810,000. The gestation period is 12 to 15 hours (Knott, et al., 2011; Vainio and Lagerspetz, 2006).

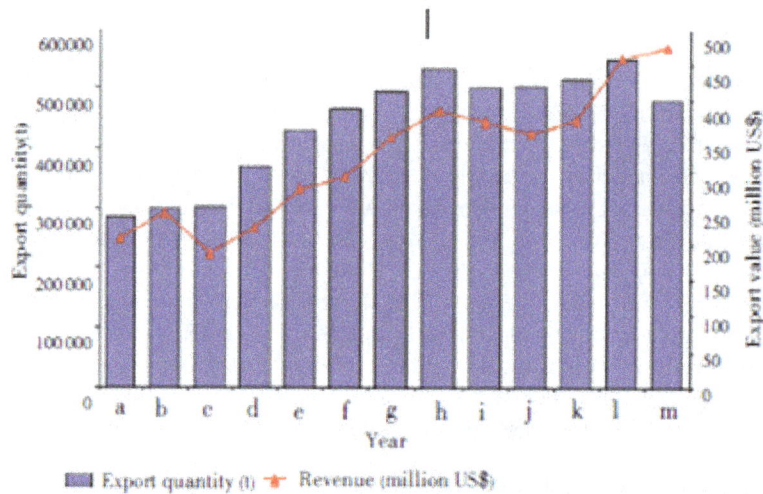

Figure 1. Export of prawn and shrimp and foreign earning in Bangladesh during 1999/2000-2011/12 (source: DoF, 2013). a: 1999-2000; b: 2000-2001; c: 2001-2002, d: 2002-2003; e: 2003-2004; f: 2004-2005; g: 2005-2006; h: 2006-2007; i: 2007-2008; j: 2008-2009; k: 2009-2010; l: 2010-2011; m: 2011-2012 (adopted from Ahamed et al., 2014).

Courtship and mating behavior of *Penaeus monodon*

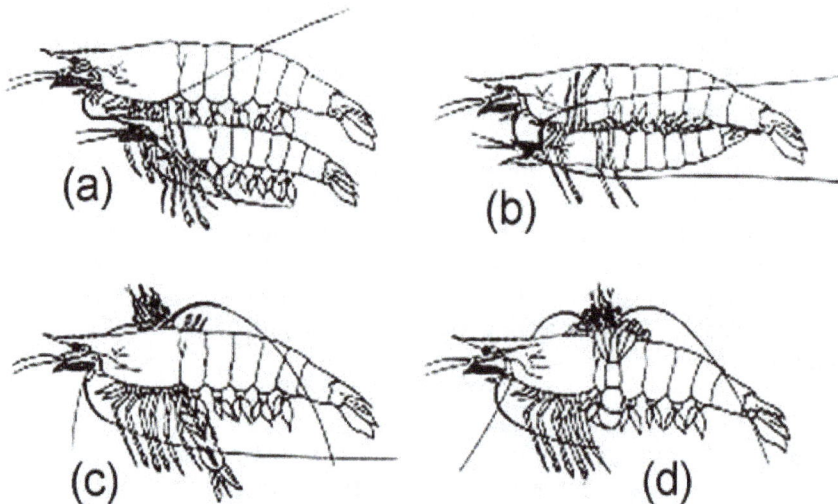

Figure 2. Courtship and mating behavior of *Penaeus monodon*

(Phase 1): A; Female above-male below in parallel swimming, (phase 2): B; Male turns ventral side up and attaches to female, (phase 3a): C; Male turns perpendicular to female (phase 3b): D; Male curves body around female and flicks head and tail simultaneously (Primavera, 1979).

Ovarian maturation

The maturation of the ovary has been categorized into five stages, the classification of which is based on ovum size, gonad expansion, and coloration.

Stage I and V (undeveloped and spent stages)

Ovaries are thin, transparent, and not visible through the dorsal exoskeleton. Stage I is known as the perinuclear stage composed of perinuclear oocytes (46-72 microns). Oocytes bigger than 55 microns are enveloped by a single layer of follicle cells. Similar features are observed in the spent stage (stag v) which also contains some yolky oocytes, thicker follicle layer, or irregularly shaped perinucleolar oocytes.

Stage II (developing stage)

The ovaries are white to olive green in color, and discernible as a linear band through the exoskeleton. The developing ova averaging 177 microns in diameter have yolk granules and cells believed to be nutritive bodies. The cells referred by as cystoplasmic inclusions are composed of small granules of glycoproteins, medium-sized globules of lipoglycoproteins, and few large lipid droplets.

Stage III (nearly ripe stage)

The anterior portion of ovaries is thick and expanded. They are very visible through the exoskeleton, particularly at the first abdominal segment, when viewed against the light. The ova average 215 microns in diameter.

Stage IV (ripe stage)

The ovary classified as ripe (mature) stage is diamond-shaped, expanding through the exoskeleton of the first abdominal segment. The isolated ovary appears dark olive green, filling up all the available space in the body cavity. There is the presence of a characteristic margin of peripheral rod-like bodies, the apexes of which radiate from the center of the egg. The ova average 235 microns in diameter (Akand et al., 1990).

Life cycle

Shrimp mature and breed only in a marine habitat. The females lay 100,000 to 500,000 eggs, which hatch after some 24 hours into tiny nauplii. These nauplii feed on yolk reserves within their bodies, and then metamorphose into zoeae. Shrimp in this second larval stage feed in the wild on algae, and after a few days, morph again into myses. The myses look akin to tiny shrimp, and feed on algae and zooplankton. After another three to four days, they metamorphose a final time into postlarvae: young shrimp that have adult characteristics. The whole process takes about 12 days from hatching. In the wild, post-larvae then migrate into estuaries, which are rich in nutrients and low in salinity. They migrate back into open waters when they mature.

Development of prawn farming

Around three-quarters of prawn farms are located in the southwest part of Bangladesh which has been identified as the most important and promising area for prawn culture, because of the availability of wild postlarvae, favorable resources and climatic conditions, such as the availability of ponds, low lying agricultural land, warm climate, fertile soil, and cheap and abundant labor (Ahmed 2001). In 2002, there were an estimated 105000 prawn farms in Bangladesh, of which 75000 (71%) were located in the southwest (Muir 2003). At that time, there were 30 000 ha of land under prawn farming (Williams 2003); this compares with 3500 ha in the mid 1980s (Muir 2003a). At present, the prawn culture area has increased to an estimated 50 000 ha (Khondaker 2007). The freshwater prawnfarming starts in Bangladesh dates back to the early 1970s in the Satkhira district (Ahmed et al. 2013). Early 1990s, the culture of these freshwater resources spread to the southwest Bagerhat district and further to other neighboring districts including Khulna and Jessore day by day. The areas ofshrimp farm were mostly concentrated in few upazila of the districts Khulna, Bagerhat, Satkhira and Cox's Bazar in early 70's. The area of shrimp farm increased very rapidly during 1980 – '90. In 1982-83, the area was around 39,496 ha, while now it is around 115,088 ha which is 3.43 times more than the past (DoF, 2007).Therefore, the beginning of the last decade has also witnessed the development of prawn farming to other parts of Bangladesh including the Noakhali, Patuakhali and Mymensingh districts. A recent study shows that around 75% of the prawn farms are still located in the southwest part of Bangladesh

Culture technique

There have three types of culture system in Bangladesh. In extensive culture system, the shrimps are fully dependent on the availability of natural food in the pond. The amount of natural food organisms becomes insufficient as the shrimps grow. When natural food diminishes the growth of shrimps slow down. This usually occurs towards the second month of the culture period depending on the stocking density and fertility of the pond. Additional feeds should be given in the form of supplemental feeds throughout the culture period to maintain the optimum growth rate. Pond fertilization is done when the natural food in the pond is diminishing. Transferring of stock to a new pond with luxuriant growth of natural food has been found to be maximum growth of shrimp but mortality during transfer is unavoidable especially when the shrimp have newly molted. The characteristic features of improved extensive system includes low stocking density, irregular fertilizing and feeding while in semi intensive system, medium stocking density, regular fertilizer and handmade feed (sometime commercial feed) are used , water exchange are performed when need.

Pond preparation

The bottom soil plays a major role in any earthen pond culture system. Natural food organisms are one of the most important food sources in ponds. It is rich in protein, vitamins, minerals and other essential growth elements that simple supplementary feed cannot complete (Hussain and Uddin 1995). The drying of the pond bottom is the most cheap and effective method of eliminating undesirable species in pond prior to the culture period. Drying facilitates mineralization of organic matter and oxidizes harmful chemical substances. The chemicals used for liming of soils are Calcium oxide, Calcium hydroxide, Calcium $CaCo_3$ and mixed calcium-magnesium carbonate. About 69.92% farmers used lime at different doses in Bagerhat region of Bangladesh (Ahmed et al., 2008).

Stocking density

The highest stocking density of bagda PL was found in Mongla (40.95 thousand/ha) whereas lowest in Mollahat (8.08 thousand/ha) Thana. Stocking density of 15 and 22 pcs/m2 exhibited the highest production of 4635.1±128 kg/ha/crop and 4328.7±138.2 kg/ha/crop respectively (Saifullah et al., 2005). Stocking density between 10-20 pcs/m2 is ideal for successful shrimp farms and for the nursery of shrimp the stocking density of 100pcs/m2 is given better production. The fry should be stocked in the early morning (7 to 10 am) or late in the evening (9 to 11pm) when the pond water temperature is low.

Seasons of fry availability

Marketing of prawn seed starts in the month of March- April and the peak season of fry marketing is May to July. The maximum harvest from nature occurs in this time. Hatchery produced seeds are available between March to September (Saifullah et al., 2005).

Nursing of fry

Many farms use nursery ponds, tanks, cages where the post-larval shrimp are grown into juveniles for three weeks. The tanks are first filled with filtered aerated sea water. Stocking density is about 5000 fry/m^3 of water for *P. monodon*, 10,000/m^3 for *P. indicus* and *P. merguiensis*. The fry are fed with finely chopped mussel or cockle meat. Artemia nauplii are also used to minimize cannibalism. 50% of the water is changed daily. The size of nursery pond ranges from 500 to 2,000 m^2 and water depth is 40–70 cm .It is provided with at least one gate with a fine screen (1 mm mesh size) to prevent undesirable organism .Stocking density in nursery pond is about 50–150 fry/m^3 depending on the size of the fry. The nursery pond should be prepared properly. The pond is completely drained of water and dried until bottom soil cracks. Derris root at 4 g/m^3 can be applied when the pond cannot be completely drained. Lime at 500 to 2,000 kg/ha, chicken manure at 500 to 2,000 kg and inorganic fertilizer (16–20-0) at 25 to 100 kg/ha are then applied. About 30% of the water is changed daily. Chopped mussel, cockled meats are fed to the larvae at the rate of 20% total biomass. The nursing period is 30–45 days when the larvae reached 0.2–1.0 g body weight.

Nursery cages

Synthetic net cages (0.5–1.0 mm mesh size) with bamboo or wooden frames are kept afloat by bamboo raft or synthetic floats. The cages can be used to nurse shrimp larvae. Nursery cages are placed in calm water such as river, lagoon or fishpond. The cage (3 m³) is usually stocked at 1000–2000 fry/m³ of water.

Grow-out pond

In the grow-out phase, the shrimp are grown to maturity. The post-larvae are transferred to ponds where they are fed until they reach marketable size, which takes about three to six months.

Food and feeding habit

The food of *Penaeus monodon* consisted mainly of Crustacea (small crabs and shrimps) and mollusks, making up around 85% of ingested food. The remaining 15% consisted of fish, fish larvae and scales, polychaetes, ophiuroids, debris, sand, and silt. The foods contents of *P. monodon* are divided into four broad categories (Marte, 1980) .These are digested material and detritus, vegetable substance, crustaceans and non-crustaceans and food pellets. In Zoeal stage, they are herbivorous. At this stage they prefer phytoplankton such as Chaetocerosand Skeletonema. In Mysis stage (5-7 days after hatching), they become carnivorous. Their preferences shifting from phytoplankton to zooplanktons such as the Brachionus, the brine shrimp, Artemia and other zooplanktons. In postlarval stage, they started to ingest small crustaceans like crabs and shrimps, mollusks, fish, ophiuroids, polychaetes and even debris, sand and silt. They eat both plants and animals but when starved they eat any food offered to them. The staple food of *Penaeus monodon* is crustaceans though mollusks are also eaten in large amounts. In the wild environment, they prefer small shrimps and fishes. They feed on mollusks and fish for their gonad development and to attain sexual maturity (Pascual, 1989). They are opportunistic in feeding behavior. The feeding activity of female prawns is significantly higher than that of the males having same age. The adult is act as a predator especially for the slow-moving benthic macro-invertebrates (Motoh et al., 1985). They are slow eaters. They take food with their pincers then bring this to their mouth, and finally chew on the food slowly. In case of small size of the food they throw the whole piece into the mouth. Though they have been found to eat almost allthe day, they seem to eat more at night than at daytime. Before low tidethey proliferatetheir feeding activity. When the food is of poor quality and become insufficient they become cannibalistic in nature. There is also evidence that the healthy prawns attack the weak ones or those that have just molted to feed on (Ramanathan et al., 2005).

Supplemental feeds

Supplemental feeds provide nutrition in case of insufficient natural food for increased growth. Different artificial feeds consisting of cooked rice, fishmeal, oil cake, kura, chira, snail muscle etc have been used in shrimp farm. Boonyaratpalm and New (1993) obtained better production by using soybean meal, cornmeal, broken rice and rice bran. William et al. (1995) reported a production of 1,024 - 1,662 kg/ha of prawn fed with formulated diet. Tidwell et al. (1993) got suitable production using fishmeal, oilcake and rice bran (Tidwell et al., 1993). The types of feed used are: Moist/wet feeds - These are prepared using locally available ingredients. The feeds should be given fresh immediately after preparation. The commonly used feeds include the following: rice bran with trash fish, house discards, chopped toads and frogs, snails' shells crushed, mussel and clam meat, snails. The other one is dry pelleted feeds - Pelleted feeds are commercially used supplementary feeds of shrimps. The feeds should have also a longer shelf – life. Supplemental feeds may be given by broadcasting, through feeding tray or automatic machine feeder.

Feeding rate and frequency

Feeding rate and frequency are essential in maximizing conversion rate of feed to shrimp. The feed is given at 5–10% of the estimated shrimp biomass per day.The common feeding frequency adopted is 2–5 times a day. Apportioning daily feed ration several times a day improve feed conversion efficiency because it reduces feed wastage, ensures feed quality. If the stocks are to be fed 5 times a day, two should be given in daytime and 3 at night as the shrimps are more active when dark.

Recommended nutrient levels for prawn and shrimp feed on percentage fed basis

Shrimp size (gm)	Protein (%)	Fat (%)	Fiber (%)	Ash (%)	Moisture (%)	Calcium (%)	Phosphorus (%)
0.0-0.5	45	7.5	Max.4	Max.15	Max.12	Max.2.3	Min. 1.5
0.5-3.0	40	6.7	Max.4	Max.15	Max.12	Max.2.3	Min. 1.5
3.0-15.0	38	6.3	Max.4	Max.15	Max.12	Max.2.3	Min. 1.5
15.0-40.0	36	6.0	Max.4	Max.15	Max.12	Max.2.3	Min. 1.5

(Source : Lin, 1994)

Aeration and water exchange

Four aerators are needed for a 0.5-1.0 ha pond. These are installed at the corners of the pond, usually 3-5 m from the bottom of the dike. The type of aerator to be used depends on the depth of the water. One horsepower paddle wheel aerators should be used in ponds of less than 1.2 m water depth and the 2 HP paddle wheel aerators should be used in ponds deeper than 1.2 m. The most popular type of aerator is the long arm paddle wheel aerator. Change of pond water is important to maintain water quality. The process also helps to introduce new food organisms into the pond and stimulate molting of shrimp. The water in the pond can be changed through tidal flows or by means of a mechanical pump. Tidal exchange of pond water is practiced in traditional culture. Water in the pond is drained to one half of the pond level during low tide and is replenished during rising tide. Water pumps of various capacities are used to replenish pond water in semi-intensive and intensive culture operation.

Harvesting

Shrimp can be harvested in good condition within a short period of time. The harvesting should be done carefully so that it does not damage or contaminate shrimp. Rapid harvesting will reduce the risk of bacterial contamination. Complete harvesting is done by draining the pond water through. The average culture period required is around 120-150 days during which time the prawns will grow to 20-30 gm size (depending on the species). It is possible to get two crops in a year. Harvested shrimps can be kept between layers of crushed ice before transporting.

Prawn productivity

The average annual yield of head-on prawns in Bangladesh was reported to be 336 kg/ha (Muir 2003). The average productivity of prawn has increased in recent years, probably as farmers have become more confident to increase stocking densities and feeding levels. In the early1990s, the average yield of prawn was only168 kg ha-1, which was low due to the traditional farming method and the relatively low level of inputs (Rahman 1994). However, in the late 1990s, reported yields had increased, with a typical yield of 200-250 kg/ha being obtained (Rahman 1999), while Hoq, Islam and Hossain (1996) reported that prawn production when reared together with fish, varied from 162 to 428 kg ha-1. Nevertheless, most of the prawns are cultivated using extensive methods in Bangladesh and productivity is low compared with other countries. The quantity of total prawn production in Bangladesh remains rather uncertain because production statistics often do not distinguish between prawn and shrimp.

CONCLUSION

Freshwater prawn farming plays an important role in the economy of Bangladesh, earning valuable foreign exchange and contributing to increased food production, diversifying the economy and increased employment opportunities. In spite of several problems, the practice of prawn farming has offered an opportunity to increase incomes for farmers and associated groups. Shrimp farming offers significant employment opportunities, which may help alleviate the poverty of the local coastal populations in many areas, if it is properly managed. Estimates of the labor intensity of shrimp farms range from about one-third to three times more than when the same area was used for rice paddies, with much regional variation and depending on the type of farms surveyed. Despite the great potential of shrimp culture in this country, successful commercial farming is facing a number of challenges, including lack of quality seed, poor quality feed, lower production, inadequate food safety and quality control and marketing constraints. So, it is necessary to overcome these challenges with the help of institutional and organizational support, technical assistance, improved government facilities and infrastructure, extension services and training programmes.

CONFLICT OF INTEREST

The authors declare that they have no competing interests.

REFERENCES

1. Ahmed N, Demaine H, Muir JF. 2008. Freshwater prawn farming in Bangladesh: history, present status and future prospects. Aquaculture Research, 39: 806-819.

2. Ahmed N, Ambrogi AO, Muir JF. 2013. The impact of climate change on prawn postlarvae fishing in coastal Bangladesh: socioeconomic and ecological perspectives. Market Policy, 39: 224-233.

3. Akand AM, Hasan MR. 1992. Status of freshwater prawn (*Macrobrachium* spp.) culture in Bangladesh. In: Silas EG, Sebastian MJ, Thampy DM, Rabindranath P, Mathew PM, editors. Symposium series in fisheries (India), no. 1; National symposium on freshwater prawns; 1990 Dec 12-14; Kochi, India. Vellanikkara, India: Kerala Agricultural University; p. 33-41.

4. Department of Fisheries (DoF). 2013. Fishery statistical yearbook of Bangladesh 2011-2012. Dhaka, Bangladesh: DoF, Ministry of Fisheries and Livestock.

5. DoF. 2007. Annual Report (2005-2006). 27-51 pp.

6. Dev, B K. Present Status of Shrimp at the Stage of Production and Marketing: A Study in Khulna District of Bangladesh.) Fisheries Extension Officer, Department of Fisheries, Ministry of Fisheries and Livestock http://bea-bd.org/site/images/pdf/26.pdf

7. FAO. 2012 Species Fact Sheets: *Penaeus monodon* (Fabricius, 1798). In: FAO Fisheries and Aquaculture Department. http://www.fao.org.

8. Hossain M I, AM Shahabuddin, M. A. B. Bhuyain, M. A. Mannan, M. N. D. Khan and R. Ahmed. 2013. Scaling Up of Stocking Density of Tiger Shrimp (*Penaeus monodon*) under Improved Farming System in Khulna Region of Bangladesh. American Journal of Experimental Agriculture, 3: 839-848.

9. Hussain MM, Uddin MH. 1995. Quality control and marketing of fish and fish products: needs for infrastructure and legal support. Dhaka: Report of the national workshop on fisheries resources development; 1995 Oct-Nov. Report No: FAO-FI--GCP/RAS/150/ DEN.

10. Khan MSA, Alam MJ, Rahman F, Shah MMR. 2004. Optimization of stocking densities of *Penaeus monodon* post-larvae in brackish water pond. Bangladesh Journal of Zoology, 32: 101-107.

11. Kungvankij P. and T.E. Chua. 1986. Shrimp culture: pond design, operation and management. Food and Agriculture Organization of the United Nations (FAO).

12. Motoh H. 1985 Biology and ecology of *Penaeus monodon*. In: Taki Y, Primavera JH, Llobrera JA (eds), Proceedings of the First International Conference on the Culture of Penaeid Prawns/Shrimps, 4-7 December 1984, Iliolo City, Philippines. Aquaculture Department, Southeast Asian Fisheries Development Center, pp 27–36.

13. Motoh H. 1981. Studies on the fisheries biology of the giant tiger prawn, *Penaeus monodon* in the Philippines. Technical Report No. 7, SEAFDEC, Philippines, 128 pp.

14. Muir J.F. 2003. The future for Fishsheries: economic performance. Fisheries Sector Review and Future Development Study. Commissioned with the association of the World Bank, DANIDA, USAID, FAO, DFID with the cooperation of the Bangladesh Ministry of Fisheries and Livestock and the Department of Fisheries, Dhaka, 172pp.

15. Pascual FP, 1989. Nutrition and feeding of *Penaeus monodon* (3rd ed.). Tigbauan, Iloilo, Philippines: Aquaculture Department, Southeast Asian Fisheries Development Center. pp 1–23.

16. Ramanathan N, Padmavathy P, Francis T, Athithian S, Selvaranjitham N. 2005. Manual on polyculture of tiger shrimp and carps in freshwater, Tamil Nadu Veterinary and Animal Sciences University, Fisheries College and Research Institute, Thothukudi.

17. Saifullah AS, Rahman MS, Jabber SM, Khan YS, Uddin N. 2005. Study on some aspects of biology of prawns from north east and northwest regions of Bangladesh. Pakistan Journal of Biological Science, 8: 425-428.

18. Tidwell JH, Webster CD, Yancy DH and Abramo LRD, 1993. Partial and total replacement of fishmeal with soyabean meal and distillers' by products in diets for pond culture of the freshwater prawns (*Macrobrachium rosenbergii*). Aquaculture, 118: 119-130.

19. William HD, Abramo LRD, Fondren MW and Duran MD, 1995. Effects of stocking density and feed on production characteristics and revenue of harvested freshwater prawns *Macrobrachium rosenbergii* stocked as size-graded juveniles. Journal World Aquaculture Society, 26: 38-47.

19

OPTIMIZATION OF DOSES OF MEGAVIT AND AQUABOOST ON THE GROWTH PERFORMANCE OF THAI PANGAS (*Pangasius sutchi*) IN AQUARIUM CONDITIONS

Sonia Sku, Md. Golam Sajed Riar*, **Sumit Kumer Paul[1], Nur-A-Raushon and Md. Kamal[1]**

Freshwater Station, Bangladesh Fisheries Research Institute, Mymensingh-2201, Bangladesh; [1]Department of Fisheries Technology, Faculty of Fisheries, Bangladesh Agricultural University, Mymensingh-2202, Bangladesh

*Corresponding author: Md. Golam Sajed Riar; E-mail: riarsajed@yahoo.com

---

ARTICLE INFO

Key words

Pangasius sutchi
Growth promoter
Aquaboost
Megavit

ABSTRACT

The present study was conducted to optimize doses of Megavit and Aquaboost on the growth performance of Thai pangas (*Pangasius sutchi*) in aquarium conditions. Six treatment with three replications were used (i) control; (ii) with less than recommended dose (LD) of aqua drugs (0.66 g/kg of the feed); (iii) recommended dose (RD) of aqua drugs (1 g/kg of the feed) and (iv), (v) and (vi) more than recommended doses (MRD) of aqua drugs (1.33g/ kg, 1.40 g/kg and 1.50g/kg of the feed) for 28 days in aquarium condition. In each aquarium 10 fingerlings of average initial weight 9.35 g were stocked. Feed containing either Aquaboost or Megavit, fed at 10% body weight of fish two times daily. The body weight was determined with time interval. Mortality rate also recorded. In case control feed, body weight increased slightly from initial 9.35 to 10.75g in 28 days but a positive trend in growth performance of pangas was observed after feeding different doses of aquaboost medicated diets. The highest increased body weight of 13.28±0.08g was observed in the treatment (iv) containing 1.33g/kg aquaboost in the diet of more than recommended doses where in recommended dose (iii) of 1 g/kg, the body weight of pangas increased to 11.52±0.09g. However, with the more recommended doses (MRD), (iv), (v) and (vi) in the diets, growth performance was declined. Similar results were also observed when feeding Megavit to pangas in 28 days in aquarium.

INTRODUCTION

Aquaculture in Bangladesh is an important economic activity to the farmers and entrepreneurs. Now a day, pangas is an important fish species grown widely in our aquaculture system. Depending on the availability of fry the pangas has widely been spread throughout the country and become one of the most popular cultivable species due to its high yield and low production cost. It is now widely believed that various types of aqua-medicinal products are indeed essential ingredients for successful aquaculture which has been used in various forms for centuries (Subasinghe et al., 1996). Aquaculture chemicals are widely used for health management of aquatic animal, pond construction, soil and water quality management, improving natural aquatic productivity, transportation of live fish, feed formulation, manipulation of reproduction, growth promotion, processing and value addition of the final product (GESAMP, 1997 ; Subasinghe et al., 1996). It is clearly evident that intensification of aquaculture brings about the use of more chemicals and antibiotics in this sector.

Commonly used chemicals in Bangladesh aquaculture are lime, rotenone, various forms of inorganic and organic fertilizers, phostoxin, salt, dipterex, antimicrobials, potassium permanganate, copper sulphate, formalin, sumithion, melathion, etc. (Plumb, 1992; Alderman, 1992; Phillips, 1996; Hasan and Ahmed, 2002; Brown and Brooks, 2002; DoF, 2002 ; Faruk et al., 2004.)

Different types of aqua medicinal products are widely used in our aquaculture system. There is a little work done in the past to evaluate their efficacy and safe doses for the production of cultivable fish species. It is also fact that aqua medicinal products manufacturing companies are usually marketed these products in Bangladesh but farmers didn't follow the standard rules which is given by the researcher and the company. It is important to evaluate the performances of the aqua drugs through field testing and optimize their doses. The work has been undertaken to see the effects of two aqua drugs, Aquaboost and Megavit on the growth performance of pangas (*Pangasius sutchi*) initially in aquarium conditions.

MATERIALS AND METHODS

Study area

The study was carried out in aquariums of Fisheries Technology Laboratory, Bangladesh Agricultural University (BAU) campus, Mymensingh. Two aqua drugs; Aquaboost and Megavit were used as growth promoter and its effect on the growth performance of pangas. The total duration of the experiment continued for 28 days in aquarium conditions. The details of the experimental conditions are given below:

Experimental conditions in aquarium:

Megavit and Aquaboost, both chemicals produced by Novartis were used to evaluate the effective doses for the growth performance of pangas *(Pangasius sutchi)*.Six treatments with three replicates were used in aquarium condition. Ten fingerlings (9.35g average body weight) were stocked and fed at 10% body weight two times daily in each aquarium. Total water holding capacity of individual aquarium was 50 litres. All the aquaria were covered by a bamboo made net and aeration was provided with aerator. Water quality parameters such as pH, Temperature, alkalinity, hardness, NO3, ammonia and DO were examined properly.

Feed Formulation

Pelleted feed containing either Aquaboost or Megavit were prepared with rice husks (58.33%), maize flour (6.67%), mustard oilcake (15%), meat bone (15%) and dry fish meal (5%). These feeds were divided into six groups: (i) control (feed was prepared without Aquaboost or Megavit), (ii) 0.66g/kg of feed (less than recommended dose) (iii) 1g/kg feed (recommended dose) (iv) 1.33g/kg (more than recommended dose), (v) 1.40g/kg (more than recommended dose) and (vi) 1.50g/kg (more than recommended dose), respectively. Feeds were dried and stored in refrigerated temperature 4°C. Feeds were prepared in Tora agro farm located Dhanikola of Trisal upazilla, Mymensingh. The proximate composition of applied feed was determined according to standard procedure.

Analysis of proximate composition of feeds

Proximate composition such as Moisture, Ash, Crude Protein, Crude Lipid and Crude Fibre of prepared feed was determined according to the standard methods given in AOAC (1980).

Specific Growth Rate (SGR)

The specific growth rate (SGR) was determined by using following formula:

SGR (% day) = [log W2- log W1] /T2- T1.

Here,

W2= Mean final weight (g), W1 = Mean initial weight (g),

T2 = Time at end of the experiment, T1 = Time at initial of the experiment.

Statistical analysis

The data obtained in the experiment were analyzed by using ANOVA. The mean values compared using Duncan's Multiple Ranged Test (DMRT) to identify the level of significance of variance among the treatments as post-hoc test using SPSS (Statistical Package for Social Science, version 11.5) statistical software (SPSS mc; Chicago. USA). Significant differences were determined among treatments at the 5 % level ($P < 0.05$).

RESULTS

Proximate composition of feed

The result shows that moisture content in different medicated feeds and control feeds were more or less similar with little variation. The moisture content was in the range of 12.36 to 16.24%. The lipid content in the feed was in the range 7.70 to 8.90%, crude protein in the range of 23.10 to 24.68% and ash content in the range of 12.02 to 13.72%. There is little or no variation in crude fibre content which is within 7%. Carbohydrate was in the range of 32 to 37%. The composition of feeds in the present study is little lower than the acceptable limit for catfish where protein requirement is 25–50% (Robinson et al 2011).

Growth performance of pangas (*Pangasius sutchi*) after feeding different doses of Aquaboost tread feed has been presented in Table 2. After feeding control and medicated diets, the growth performance in different treatments were evaluated with time interval by taking sample weight from each treatment. In case control feed, body weight increased finally from 9.35g to 10.75g and a positive trend in growth performance was observed in other treatments after feeding different doses of medicated diets. The highest body weight increased from 9.35 g to 13.28±0.08g was observed in the treatment (iv) containing 1.33g/kg. The body weight increase in treatment (iii) containing recommended dose of 1 g/kg was 11.52±0.09g. The body weight declined considerably in case of more than recommended doses in the treatment (v) and (vi) (1.40g/kg and 1.50g/kg) over 1.33g/kg. The results indicate that 1.33g/kg Aquaboost in the diet might be safe dose for the pangas culture.

Table 1. Proximate composition of feed (*= The values within parentheses are calculated based on dry weight basis)

Name of item	Moisture (%)	Lipid (%)	Crude protein (%)	Ash (%)	Crude fiber (%)	Carbohydrate (%)
Control feed	12.36	7.70 (8.78)*	23.10 (26.35)*	13.72 (15.65)*	6.56 (7.58)*	36.56 (41.72)*
Aquaboost (0.066 g/kg – 1.50 g/kg)	12.98 - 13.17	8.56 (9.85)* ⁻8.90 (10.22)*	24.50 (28.21)* ⁻24.68 (28.34)*	12.03 (13.85)* ⁻12.02 (14.35)*	6.20 (7.14)* ⁻6.30 (7.23)*	35.54 (40.93)* ⁻35.23 (40.45)*
Megavit (0.066 g/kg – 1.50 g/kg)	16.24 - 15.16	8.60 (10.26)* ⁻8.60 (10.13)*	24.15 (28.83)* ⁻23.98 (28.26)*	12.02 (14.35)* ⁻12.07 (14.22)*	6.80 (8.11)* ⁻6.50 (7.66)*	32.19 (38.43)* ⁻33.69 (39.71)*

*Chemical composition of feeds treated with different doses of either Aquaboost or Megavit and control feeds used in the experiment

Fisheries Science: A Global Assessment

Table 2. Growth performance of pangas (*Pangasius sutchi*) feeding different doses of Aquaboost tread farm made feed in Aquarium condition

Aquarium	Weight of fish (g)				
Doses	0 day	7 days	14 days	21 days	28 days
Control	9.63±0.19	7.25±0.19	10.10±0.16	10.25±0.19	10.57±0.12
0.66 g/kg	9.35±0.11	9.64±0.12	10.14±0.13	10.18±0.11	10.40±0.09
1.0 g/kg (RD)	9.54±0.07	9.98±0.09	10.12±0.09	10.22±0.07	11.52±0.09
1.33 g/kg	9.64±0.10	10.23±0.08	10.73±0.11	11.43±0.10	13.28±0.08
1.40 g/kg	9.45±0.13	9.68±0.13	10.24±0.13	10.19±0.12	10.40±0.11
1.50 g/kg	9.55±0.07	9.76±0.11	10.11±0.08	10.12±0.06	11.49±0.13

Table 3. Growth performance of pangas (*Pangasius sutchi*) feeding different doses of Megavit tread farm made feed in Aquarium condition

Aquarium	Weight of fish (g)				
Dose	Initial avg. (g)	7 days (g)	14 days (g)	21 days (g)	28 days (g)
Control(g/kg)	9.63±0.19	7.56±0.19	10.10±0.19	10.25±0.16	10.57±0.12
0.66 g/kg	9.49±0.13	8.92±0.16	10.26±0.15	10.53±0.13	10.83±0.16
1.0 g/kg (RD)	9.44±0.13	9.49±0.15	10.18±0.13	10.42±0.13	11.95±0.15
1.33 g/kg	9.345±0.13	7.56±0.09	10.83±0.11	11.66±0.13	12.50±0.09
1.40 g/kg	9.47±0.14	9.92±0.14	10.28±0.17	10.43±0.12	11.28±0.13
1.50 g/kg	9.47±0.11	9.59±0.13	10.17±0.14	10.44±0.15	11.55±0.12

Table 4. Water quality parameter of aquarium

Parameter	0 Days	After 7 Days	After 14 Days	After 21 Days	After 28 Days
Temperature(^{0}C)	28.0 ± 0.47	29.7 ± 0.79	30.4 ± 1.30	30.5 ± 0.44	31 ± 0.58
pH	8.50 ± 0.38	8.35 ± 0.13	8.45 ± 0.07	7.6 ± 0.15	7.9 ± 0.12
Alkalinity	120 ± 3.06	100 ± 5.77	150 ± 2.87	120 ± 15	100 ± 4.58
DO	5 ± 0.29	5.5 ± 0.15	6.5 ± 0.09	7 ± 0.06	6 ± 0.06
Hardness	50 ± 2.08	200 ± 6.01	100 ± 5.21	150 ± 4.41	200 ± 2.89
Nitrite	0.03 ± 0	00	0.03 ±00	00	00
Ammonia	2 ± 0.12	2 ± 0.52	0.2 ± 0.05	0.5 ± 0.09	00

Studies were also conducted to optimize the growth performance of Megavit treated feed in aquarium conditions (Table 3). Megavit was added to the feed at different doses similarly that of Aquaboost. Samples weight was taken with time interval after feeding the control and Megavit medicated diets. A maximum body weight increase of 10.80g was obtained in control feed in 28 days. A positive trend of growth performances was also observed in all treatments where fish were fed different doses of medicated diet. A maximum growth of 12.50±0.09 g was obtained in treatment (iv) containing 1.33g/kg in feed. A maximum increased body weight in recommended doses was found 11.95±0.15 g. The growth performance declined in other treatments containing Megavit over 1.33g/kg feed.

20

FISHERS' ACCESS TO THE COMMON PROPERTY WATERBODIES IN THE NORTHERN REGION OF BANGLADESH

Md. Amzad Hossain[1*], Mousumi Das[1], Md. Shahanoor Alam[2] and Md. Enamul Haque[3]

[1]Department of Aquaculture, Bangabandhu Sheikh Mujibur Rahman Agricultural University (BSMRAU), Gazipur-1706, Bangladesh; [2]Department of Genetics and Fish Breeding, BSMRAU, Gazipur-1706, Bangladesh; [3]Department of Agriculture Extension and Rural development, BSMRAU, Gazipur-1706, Bangladesh

*Corresponding author: Md. Amzad Hossain; E-mail: amzad@bsmrau.edu.bd

---

ARTICLE INFO

ABSTRACT

Key words
Fishers
Access
Common property
waterbody
Biodiversity

The study was designed to explore the status of fishers' access to the common property waterbodies (CPW) and associated problems of using CPW. Three upazillas (administrative units) of the Northern region of Bangladesh were selected for the empirical study. Data were collected from fishers, non-fishers and other stakeholders through structured interview schedules, physical observation, and participatory rural appraisal (PRA). The access of poor fishers group to the CPW was very limited in the study area. The government policy of revenue collection through leasing system badly affected the fishers group as they do not have required level of capital, unity, leadership and education. A revenue oriented fisheries management system with short lease periods was found to encourage over-fishing and destructive fishing by lessees, where the lessees were noted to sweep away all the fish stock as soon as their contract ended without considering the sustainable use of resource and biodiversity. Consequently, the productivity of the CPW is declining gradually. To ensure the effective access of fishers' group to CPW and their sustainable use, an advised long-term community based management (CBM) plan needs to be developed with the effective participation of the fishers' groups and other stakeholders.

INTRODUCTION

Fisheries are important sub-sector of agriculture in Bangladesh and play a significant role in nutrition, employment, foreign exchange earnings and food supply (Dey et al., 2005; Roos et al., 2007). The fishery sub-sector contributed 4.37% to GDP at current price during 2012-13. Around 16.2 million people earn their livelihood directly or indirectly from activities related to fisheries (DoF, 2014). Total fish production in 2012-2013 was 3.41 million Metric Tons (MT) of which 2.82 million MT came from inland waterbodies and 0.59 million MT from marine fisheries (FRSS, 2014).

Historically, most of inland waterbodies were non-private or state property where fishing, animal grazing, fodder and plant harvesting were open to all. Those waterbodies are known as common property waterbodies (CPW). In course of time the government took away free fishing rights from the relatively large waterbodies (e.g., rivers, beels and haors) through establishment of periodic leasing system in order to generate revenues (Toufique, 1999). To facilitate the leasing process rivers and their tributaries are divided by the Ministry of Land (MoL) into several small arbitrary segments. These segments or waterbodies are then leased out through auction for the collection of revenue. Similarly, beels (land depressions) and ponds owned by the government fall under this category. There are over 10,000 waterbodies (inland waterbodies generating government revenue) in Bangladesh (Viswanathan et al., 2002) and they are leased to the highest bidder with a preference for fisher cooperatives but very often, either directly or by bidding through a cooperative, control ends in the hands of the rich and influential lessees. Due to the private auction leasing system, fishers' access to inland fisheries has become increasingly difficult and competition over the fisheries resources is becoming more intense and complex every year. This leasing system created a group of middle agents, usually from rich and elite class, who with their economic and social powers, established perpetual authority over these resources and continued to be benefited at the expense of the professional and hereditary fishers (Khan, 2012).

Over the years, introduction of new agricultural technologies and increased competitive pressures for more intensive use of the wetlands resulted in the deterioration of fisheries resources as well as reduced the size of CPW. Nevertheless, traditional rules of fishing rights by the community were maintained, at least in the water areas with marginal or no use value for agricultural purposes. Development policies with respect to these water resources seldom consider the interests of the poor communities who used to derive benefits for their livelihood (Ahmed, 1993).

In the above context, it is evident that the present status of CPW in economically deprived areas, their access to poor fishers' community and most importantly their management strategy need to be clarified to satisfy their sustainable use. Therefore, in this study a pragmatic approach was undertaken to explore the empirical picture of fishers' access to the CPW and their provident consequences in adversely poverty affected Northern region of Bangladesh.

MATERIALS AND METHODS

Study area and sampling frame

In the present study government owned waterbodies, which can be accessed or should be accessed by common people free of cost or on lease basis, are considered as CPW (river, flood plain, beel, pond etc.). Data were collected from villages adjacent to the selected CPW under three upazillas (administrative unit) of the northern region of Bangladesh (Fig.1). Details of the sampling design are shown in Table 1.

Table 1. Details of the sampling design

District	Upazilla	Selected CPW	Village	Fisher respondent	Non-Fisher respondent
Lalmonirhat	Lalmonirhat sadar	Ratnai Nodi	Kulaghat, Khatamari	106	51
Kurigram	Kurigram sadar	Dasherhat Charra	Polasbari, Cherenga	95	71
Rangpur	Pirgaccha	Masankura Moranodi	Nijpara, Kabila para	73	56
		Total		**274**	**178**

Data collection methods

The study was conducted through Participatory Rural Appraisal (PRA), survey, monitoring, discussion and consultation among the resource users and stakeholders. Two interview schedules were used for data collection. One was for fishers, non-fishers, and another for different stakeholders including upazilla and district level Personnel of MoL, Department of Fisheries and Ministry of Youth and Sports. Fishers and non-fishers in the study area were selected through simple random sampling method. Fishermen were interviewed at home and or fishing sites. In a given day approximately five to seven interviews were conducted where each interview schedule for fishermen was addressed the issue of fishing activities, access to CPW and their problems relating to access.

PRA tool such as focus group discussion (FGD) was done with fisher and non-fisher groups in the study areas. Nine FGD sessions, three from each location, were conducted, where each group had 10 to 14 people. FGD sessions were held in front of village shops and on the bank of CPW. Secondary data were collected from relevant upazilla and district level Govt. offices, statistical yearbook, project reports, scientific articles and websites.

Data processing and analysis

Data from various sources were coded and entered into database system using Microsoft excel and analyzed though simple statistical methods.

RESULTS

Present status of CPW in the study areas

The present status of CPW of three upazillas is presented in Table 2. Data show that most of the waterbodies were leased out to the cooperative in Pirgacha upazilla where in most cases the lessees were non-fishers. In Lalmonirhat and Kurigram sadar the huge percentage of non-leased waterbodies were comprised with river where leasing was unmanageable. Production performance of CPW was about two third in Lalmonirhat sadar comparing with other two upazillas.

Leasing system of CPW

Government declared the pro-fishers oriented Jolmohal Policies restrict the lease of CPW within fishers.If an organization of real fishers is registered with cooperative department or department of social welfare at local level then it will be qualified to participate in the lease process. But if the organization has any non-fisher member then it would not be qualified to get the Jolmohal settlement. Individual person or unregistered Community Based Organization (CBO) does not qualify to apply for CPW. Some important features of the new policy are shown in Table 3.

Access to CPW

Fishers always did not get access to adjacent waterbodies. Most of the fisher had access to non-leased waterbodies (river and flood plain), which was about 94.5% of the total fisher, followed by beel (Fig. 2). They had access to ponds, small ditches, canals and irrigation canals in small scales. When access to the nearest waterbodies becomes restricted due to lease out or Govt. ban, the fishers have no way but to move to other open waterbodies which often far away from their locality. However, as the non-fishers and subsistence fishers do not depend on the fish catch for their livelihood, they do not go over long distance. The professional fishers go for fishing in different seasonal waterbodies. In the wet season when some low agricultural land is inundated by water they turn into seasonal fishing grounds.

These temporary fishing grounds are lost when Aman rice cultivation resumes. They exist for three to four months: from the middle of June to the middle of October. But increasingly the owners of the land where these seasonal fishing grounds are formed do not allow the fishers to fish. Some of them have started to charge a fee for access rights while others are having fish aquaculture there by themselves. The professional fishers also fish in the river for two months during March and April. In other season they are hired on contract basis for harvesting fish from the ponds during dry season. Search costs are high, as they have to move from one waterbody to another on foot. A large part of their time is spent on travelling to these marginal waterbodies. As less time available for actual fishing their income fall significantly. This exodus of the fishers to the marginal fishing grounds prompted the landowners to impose new conditions for fishing. Fishers now have to pay more for access rights to these fishing grounds.

Table 2. Present status of CPW in the study areas

	Name of Upazilla		
	Pirgacha	**Lalmonirhat sadar**	**Kurigram sadar**
Total area of CPW (ha)	71.51	282.84	12423.90
% Leased area	98.07	13.58	1.47
% Not leased area	1.93	86.42	98.53
Production rate of CPW (kg ha^{-1} year^{-1})	394.63	240.19	395.80

Table 3. Government Jolmohal Policy 2009

Criteria	Type of Jolmohal		Remarks
	CPW over 20 acres	**CPW below 20 acres**	
Eligibility	Registered Community Based Organization (CBO)	Registered CBO	All members must be fishers
Leasing authority	District Jolmohal Management Committee on behalf of MoL (DC[1] convener, RDC[2] Member Secretary)	Upazila Jolmohal Management Committee MoL (UNO[3] convener, AC[4] Land Member Secretary)	In absence of AC (Land) upazilla cooperative officer
Lease period	Max. 3 years	Max. 3 years	
Lease value	Average of last 3 years plus 5%	Average of last 3 years plus 5%	
Access	No CBO will get tenure of more than two Jolmohals	No CBO will get tenure of more than two Jolmohals	

[1]District Commissioner; [2]Revenue Deputy Commissioner; [3]Upazila Nirbahi Officer; [4]Assistant Commissioner

Table 4. Problems of using CPW

Problem	Rank
Decrease in catch from open waters	1
Short tenure of lease	2
Development of culture based fisheries in private owned seasonal flood plain	3
Control of CPW by influential person	4
Lack of alternative job in ban or lean period	5
Difficulty in forming CBO	6
Shortage of lease money	7
High lease value	8
Limited contact with concerned Govt. Official	9
Social conflict	10

Mode of involvement in fishing

Among the fishers, 85.0% were professional, they completely depend on fish capture and selling for their livelihood (Fig. 3) while 12% and rest were seasonal and subsistence respectively. The seasonal fishers generally caught fish only in the peak season of fishing but in the lean fishing season they had to find out alternative way for livelihood, as the return from fishing was very poor to ensure their living. On the other hand, majorityof non-fishers (96%) used to catch fish from CPW only for their own consumption.

Problems of using CPW

Respondents were asked about the problems of using CPW and the responses were ranked according to the prioritization (Table 4).

DISCUSSION

In the present study, 'Fisher' is considered as someone who catches fish from natural source and sells it for the livelihood. All others rural people who do not fish for their business purpose but may occasionally catch fish for their own consumption are grouped as non-fisher. There are several groups involved in fishing or fishing related activities in inland open waters of Bangladesh and they include the traditional caste fishers (mostly Hindus), non-traditional fishers (who entered fishing later), the leaseholders of waterbodies (who are mostly non-fishers), and the general fishers (members of the public) (Blanchet, 1993).

The present access status and leasing system of CPW has created substantive problems for the fishers group as well as the whole community. Decrease of fish catch from open water was the most mentioned problem by the fishers. In the past, waterbodies were full of fish. However, availability of fish has declined in the recent time. Open access to the non-leased waterbodies was the main cause for dwindling fish stock and bio-diversity. Fishers always do not follow fishing regulation and indiscriminately harvest small fish and brood fish where they have full access.

On the other hand, in case of lease out waterbodies the present Jolmohal Policy allows three years of lease period while the bureaucratic systems may take time and make the lease period even shorter. When a lessee takes lease of a waterbody for short time, he cannot contribute to improve the habitat and fish stock. On the contrary, he exploits heavily by harvesting fry, brood fishes and draining up the fish pocket or shelter in dry season before ending his tenure. Khaled (1985) examined leasing methods of river fisheries and found that overexploitation of the fisheries is encouraged by the government through its existing leasing system. Under this system leaseholders receive short-term leases with no guarantee that a lessee will be able to renew the lease of the same fishery in successive years. Barr and Dixon's (2001) studied on the management of CPR in Bangladesh and revealed that a revenue oriented fisheries management system with short lease terms where lease values increase yearly with no consideration for the productivity of waterbodies encourages over fishing and destructive fishing. The lessees most often dewater waterbodies to maximize profit at the expense of the sustainability of fish biodiversity.

Similarly, the difficulty in establishing user rights when combined with the disincentive effect of short-term leases, further reduces the return from stocking or semi-intensive aquaculture in CPW. When lake fishing shifts from capture to semi-intensive (stocking, without fertilizer use) or intensive (stocking with fertilizer use) some infrastructures are needed. Landing platforms are needed with connections to the main roads connecting to the markets, to be able to carry at a reasonable cost the high volume of fish to the market. Govt. has leased out the CPW for better management and revenue collection. However, the poor fishers' community cannot often arrange lease money and lose the control over CPW.

Recent expansion of aquaculture has reduced the access of the poor to CPW. However, when the floodplain aquaculture is conducted in private land, the landowner can only be a member. In case of Govt. land the people who can subscribe can take part in the process of fish culture. Blanchet (1993) in her study of Shanir Haor in the wetland region of Bangladesh showed how property rights, fishers access to fisheries and local fishing practices differ from the text of the law. The powerful leaseholders of water estates claim ownership over the fish stock at all times of the year. Lack of alternative job in ban or lean period of the year was another problem faced by the fishers in the study areas. When CBO adopt restrictions on fishing during the spawning season or ban on fine mesh nets, this is likely to reduce the income of fishers who depend day to day on fishing for their livelihood.

In the study area most of the respondents emphasized on the community based approach to get access to CPW and sustainable management. However, they found establishment of CBO was very difficult. This is due to lack of capital, education and integrity where influential member of the CBO often exploits them. Sometimes CBO leaders work for the interest of outsiders influential non-fishers. There were several initiatives in the country for community based management of CPW including Community Based Fisheries Management Project -1 (CBFM-1) and CBFM-2. It was found that most of the CBO formed by them worked effectively during the project period when the project gives support to the CBO. After completion of project period CBO do not work properly.

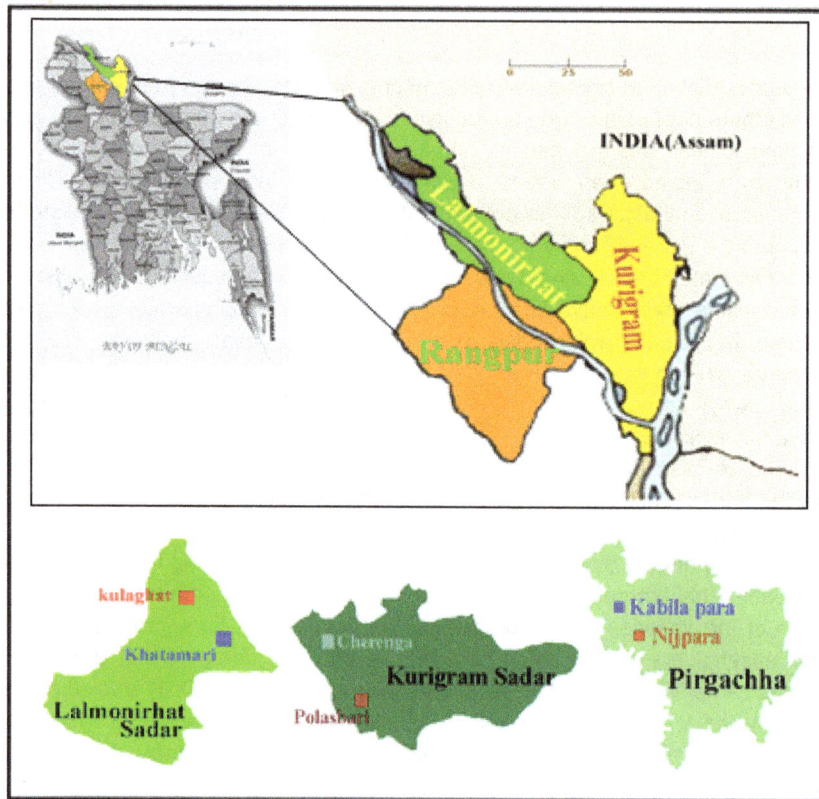

Figure 1. Map of the study area showing northern districts (sampled villages with legend)

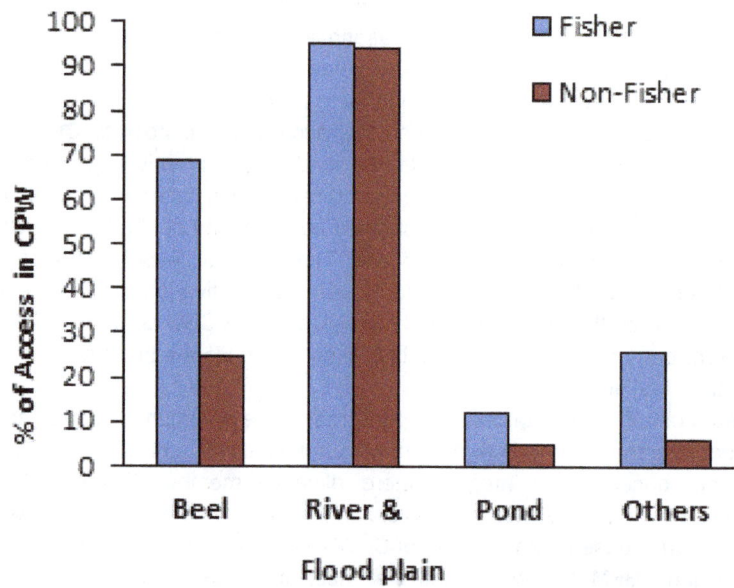

Figure 2. Present access status of CPW of fisher and non-fisher group (n = 452)

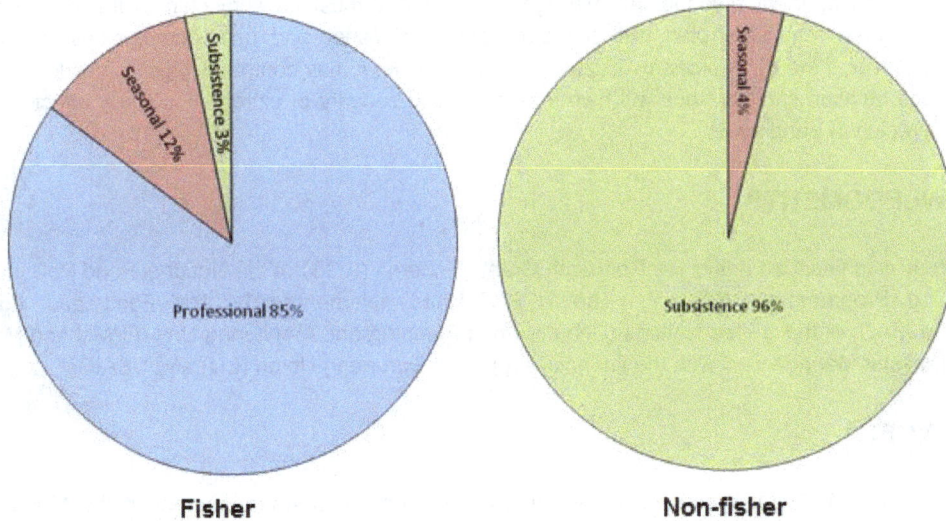

Figure 3. Mode of involvement in fishing in CPW by fishers (n = 274) and non-fishers (n = 178)

Access of fishers to CPW can be sustained through CBFM approach. CBFM controls the resource with the involvement of some government or other non-government organizations (NGO), at least for a certain period. The coastal marine fisheries resources management in Fiji, Solomon Island (Baines, 1989), coastal Japan (Ruddle, 1989), Java/Indonesia, West Africa (Lasserre and Ruddle, 1983), Mali (Moorehead, 1989) and Hawaii (Costa-Pierce, 1987; Berkes, 1996) has been shown to be successful through CBO. The Maine lobster fishery is an example of both communal and state property, where fishermen use it as a communal resource but the state maintains some management jurisdiction (Acheson, 1989). Experiences of the last decades have indicated that initiatives to alleviate poverty and achieve food security can seldom be preserved if planned without the involvement of the community. Community-centred approaches (CCA) encourage self-reliance, self-help and by doing so, raise self-esteem. Such approaches aim at empowering communities to make optimal use of locally available resources, and to effectively demand additional resources and better services to improve their livelihoods. Building on traditional social networks of support and mutual assistance, CCA mobilize community members in activities to meet their perceived needs and development priorities, thus making a significant contribution to sustainable development at local and national levels. CCA help to ensure that a range of stakeholders including women and marginal groups becomes part of the development process, real issues and needs are addressed, implementation and monitoring are improved, and sustainability enhances by giving users the leading role in developing and adapting activities. To reduce the risk of low compliance or seasonal loss of fishing incomes, Govt., local NGO should identify potentially profitable income generating activities that can compensate for restrictions on fishing and provide micro-credit and training in these activities to groups of poor fishers.

In Japan, the fishers' organization provide fund to Govt. mariculture corporations to stock coastal area with hatchery-produced fry. The fishers stop total harvesting from the stocked area or stop harvesting of the particular species for a certain period. After the self-imposed ban period they get a handsome harvest (Ruddle, 1989). This technique could be a good exemplary to manage the sustainable access of the fishers' to CPW.

CONCLUSION

Fishers could not benefit from Government policy regarding CPW and the new policy could not ensure leasing access due to lack of education, capital, unity, and leadership. In the present situation of Bangladesh complete open access to CPW was found unproductive. In open access system, fish stock declined drastically due to illegal and over fishing. Moreover, the current leasing system was found ineffective. Leasing of CPW should primarily be a means of controlling access to waterbodies to ensure sustainable management and not a system to raise government revenue. All management of CPW must be subjected to the preparation and implementation of a regulatory plan with the participation of fisher and user communities through CBM strategies. However, if the CPW properly utilized by the poor fisher it may contribute significantly to their income generation and nutrition security, thus will help to ensure food security of extremely poverty affected areas at the Northern region of Bangladesh.

ACKNOWLEDGMENTS

This study was financed under the Research Grants Scheme (RGS) of the National Food Policy Capacity Strengthening Programme (NFPCSP). The NFPCSP is implemented by the Food and Agriculture Organization (FAO) of the United Nations (UN) the Food Planning and Monitoring Unit (FPMU and Ministry of Food and Disaster Management with the financial support of European Union (EU) and USAID.

REFERENCES

1. Acheson JM, 1989. Where have all the exploiters gone? Co-management of the Maine Lobster industry. In Common property resource: ecology and community-based sustainable development, Eds., Berkes, F. Belhaven Press, London, pp: 199-217.
2. Ahmed M, 1993. Rights, Benefits and Social Justice: Keeping Common Property Freshwater Wetland Ecosystem of Bangladesh Common. Paper presented at the Fourth Common Property Conference, ICLARM, Philippines.
3. Baines GBK, 1989. Traditional resource management in the Melanesian South Pacific: a development dilemma. In Common property resources: ecology and community based sustainable development, Eds., Berkes, F. Belhaven Press, London, pp: 273-295.
4. Barr JJF and PJ Dixon, 2001. Methods for consensus building for management of common property resources. Final Technical Report for project R7562. London: DFID.
5. Berkes F, 1996. Social Systems, Ecological Systems and Property Rights. In Ecological, Economic, Cultural, and Political Principles of Institutions for the Environment, Eds., Hanna, SS, C, Folke and KG, Maler. Island Press, Washington, DC, pp: 87-107.
6. Blanchet T, 1993. Fisheries specialist study: draft final. FAP 6. Dhaka: Northeast Regional Water Management Project, CIDA with the Government of Bangladesh.
7. Costa-Pierce BA, 1987. Aquaculture in ancient Hawaii, Bioscience, 37: 320-330.
8. Dey MM, MA Rab, A Kumar, A Nisapa and M Ahmed, 2005. Food safety standard and regulatory measures: Implications for selected fish exporting Asian countries, Aquaculture Economics and Management, 9 (1&2): 217-236.
9. DoF (Department of Fisheries), 2014. Saranica, Matsya Pakhya Sankalan, Ministry of Fisheries and Livestock. The Government of Peoples republic of Bangladesh, Dhaka, Bangladesh. pp. 124.
10. FRSS, 2014. Fisheries Statistical Year Book. Fisheries Resources Survey Systems. Department of Fisheries, Ministry of Fisheries and Livestock, Dhaka, Bangladesh. pp. 50.
11. Hardin G, 1998. "Extensions of 'The tragedy of the commons'". Science, 280 (5364): 682-683.
12. Khaled MS, 1985. Production technology of the riverine fisheries in Bangladesh. In Small-scale fisheries in Asia: socio-economic analysis and policy, Eds., Theodore, P. International Development Research Centre, Canada.
13. Khan MA, MF Alam and KJ Islam, 2012. The impact of co-management on household income and expenditure: An empirical analysis of common property fishery resource management in Bangladesh. Ocean & Coastal Management, 65: 67-78.
14. Lasserre D and K Ruddle, 1983. Traditional knowledge and management of marine coastal systems. Biology International Special Issue 4.
15. Moorehead R, 1989. Changes taking places in common property resource management in the Inland Niger Delta of Mali. In Common property resources: ecology and community based sustainable development, Eds., Berkes, F. Belhaven Press, London. pp. 256-272.
16. Roos N, AM Wahab, C Chamnan and SH, Thilsted, 2007. The role of fish in food-based strategies to combat vitamin A and mineral deficiencies in developing countries. Journal of Nutrition, 137: 1106-1109.

17. Ruddle K, 1989. Solving the common-property dilemma: village fisheries rights in Japanese coastal waters. In Common property resources: ecology and community based sustainable development, Eds., Berkes, F. Belhaven Press, London. pp. 168-184.

18. Toufique KA, 1999. Property rights and power structure in inland fisheries in Bangladesh. In Sustainable inland fisheries management in Bangladesh, Eds., Middendorp, HAJ. PM. Thompson and RS. Pomeroy. ICLARM, Manila, pp. 57-63.

19. Viswanathan KK, A Mahfuzuddin, P Thompson, P Sultana, M Dey and M Torell, 2002. ICLARM-The World Fish Centres Experience with Social Research on Governance and Collective Action in Aquatic Resources, Paper presented for the conference on the Role of Social Research in CGIAR . Supporting the strategy-Achieving Development Impact, 11-14 September 2002, CIAT, Cali, Columbia.

21

EFFECTS OF ENDOPARASITISM OF *Heteropneustes fossilis* ON CONDITION FACTOR, GONAD AND BLOOD COMPOSITION OF HOST

Sarker M. Ibrahim Khalil[3], Kirtunia Juran Chandra[2] and David Rintu Das[3*]

[1]Department of Fish Health Management, Sylhet Agricultural University, Sylhet-3100 Bangladesh; [2]Department of Aquaculture, Bangladesh Agricultural University, Mymensingh-2202, Bangladesh; [3]Bangladesh Fisheries Research Institute, Freshwater Sub Station, Saidpur, Nilphamari, Bangladesh

*Corresponding author: David Rintu Das, E-mail: drd4272@yahoo.com

---

ARTICLE INFO

Key words:

Endoparasitism
Heteropneutes fossilis
Gonad
Condition factor
Blood parameter

ABSTRACT

Effects of endoparasitism of *Heteropneustes fossilis* on condition factor, gonad and blood composition of the host was conducted during July 2012 to June 2013. Fish samples were collected from various water bodies of Sylhet region. Total length, body weight, gonad weight and sex of hosts were recorded. Blood sample was collected from each fish. Six different species of parasites were identified from the hosts assessed as i). *Euclinostomum multicaecum*, ii) *Allocreadium handiai*, iii). *Lytocestus indicus*, iv). *Pseudocaryophyllaeus heteropneustus*, v). *Procamallanus heteropneustus* and vi). *Paracamallanus equispiculus*. Gonad weight, condition factors and blood parameters of *H. fossilis* were greatly affected with moderate infestations. The highest condition factor, 0.59, was found in uninfested fish and lowest, 0.45, in infested fish. Changes in the percent of haemoglobin and erythrocyte sedimentation rate (ESR) was also investigated. Percent loss of haemoglobin was 0.43 and erythrocyte sedimentation rate (ESR) was 5.19 mm/h. In general the health condition parameter due to effects of endoparasitism of the investigated fish was found negative impact for the culture production of the Singhi (*Heteropneustes fossilis*).

INTRODUCTION

Stinging catfish, popularly known as Singhi, *Heteropneustes fossilis* is an important fish in our country and getting increasingly popular showing a promising future for commercial culture (Barua, 1989). It has been contributed greatly as a delicious food fish of our country and esteemed as food for convalescence and invalids (Bhuiyan, 1964). However, once easily available in the nature, the fish has in the recent time, become scarce for many reasons including diseases and parasitic infestations (Kabata, 1985). Before starting a fish culture scheme everyone should be careful of problems, which threaten fish population and habitats. Among them effects of parasitic infestations resulting in diseases is the integral part of the existence of all animals including both cultured and wild fish population. Cross (1993) showed that normal growth of fishes is interrupted or inhibited if they are heavily infested with endoparasites.

Lots of research have been conducted on fish parasites in many countries of the world. Dias et al. (2006) studied on ecology of *Clinostomum complanatum* and recorded that the season, habitat and sex were not related to its prevalence. Hechinger et al. (2007) investigated the relationship between species richness and the abundance of larval trematodes. Mackiewicz (1982) stated that *Clarias batrachus* (Linn.) and *H. fossilis* (Bloch) are the main hosts of caryophyllaeids in the Indian subcontinent. This group of cestodes are frequently found in these catfish in the intestine and stomach region of the host. In heavy infection of caryophyllaeid cestode, they can make obstruction in the intestine by mechanical means (Chandra, 1993). Unfortunately, very few works have been initiated on the effect of parasitism on the host animals particularly in freshwater fishes in Bangladesh. As *H. fossilis* is a highly infested fish by different groups of parasites, their influence on this host is an essential task for determining its successful culture practice. Though a number of works on *H. fossilis* have been conducted in Bangladesh (Chandra and Khatun 1993, Chandra 1994, Chandra and Modak 1995, Chandra et al. 1996), effect of such parasitism on fish have not yet initiated except Laboni et al. (2012) in *Clarias batrachus*. The present research work was therefore conducted to know the effects on condition factor, gonad weight and blood parameters of *H. fossilis* due to parasitic infestations.

MATERIALS AND METHODS

A total of 210 Shingi, *H. fossilis* were collected during the period from July, 2012 to June, 2013. Live and fresh fishes were collected from Bandar market supplied from different haor named as Hakaluki haor, Tanguar haors etc. in Sylhet district. After collection, they were brought to the Disease Laboratory, Department of Fish Health Management, Sylhet Agricultural University, Sylhet by polythene bags or bucket with water in live condition for investigation. Before investigation the source, total length (TL), sex and weight of the fish were recorded in a data book. The host fish were also divided into 3 length groups (12-16, 17-21 and 22-25 cm). A slit was made on ventral side near the genital pore on anal region and was opened towards the head up to the opercular region. After careful opening the stomach and the intestine were removed and put in a petridish containing water for parasitic investigation. After collection, the parasites were kept in normal saline for relaxation, flattened and fixed in F.A.A (Formaldehyde, Acetic acids, Alcohol). Some of the parasites were prepared for permanent slides.

For collecting blood, fish were caught gently in a small scoop net and transferred into a bowl containing the same water where they were acclimatized. After anesthetizing the fish with 5 ppm quinaldine (Sigma chemical Co. USA) (Hossain and Shariff, 1992), blood samples were collected from the caudal vein with a sterile disposable plastic syringe coated with 3.6% sodium citrate as an anticoagulant according to Smith et al. (1952). To avoid contamination with mucus and water the area of insertion of syringe was wiped with alcohol cotton. Collected blood was gently pushed into a sterilized small glass vial containing anticoagulant, (potassium salt of ethylenediamine tetra-acetic acid, EDTA) to give a final concentration of 5 mg EDTA per cm^3 blood. This blood samples were used for determining the erythrocyte sedimentation rate (ESR), and haemoglobin concentration (Hb g %).

Data analysis

Gonad weight and Condition factor

The total length (cm) of each fish was taken from the tip of the lower jaw to the end of the lower lobe of the caudal fin. The body weight (g.) of each host were measured by electric balance. Condition factor was employed to evaluate the effect of parasite on the host. Condition factor was calculated by employing the formula $k = \frac{100 \times w}{l^3}$, Where w = the weight of the fish in grams; l = the length in centimeters. The magnitude of parasitism is indicated by the difference in k values of an infested and an uninfested fish. The gonad weight (g.) of infested and uninfested male and female were also measured by electric balance. Then the loss of gonad weight were measured by deducting the gonad weight of infested male and female from uninfested host.

Haemoglobin Concentration

Haemoglobin concentration was measured by the haematin method (Hesser, 1960). Sahli haemoglobinometer (Resistance LW, Germany) was used for determining the haemoglobin concentration. Prepared O.1N hydrochloric acid (HCl) was placed up to the 20-marked graduation into the perfectly cleaned and dried haemoglobinometer tube. Blood sample was drawn into the Sahli pipette exactly up to 20 cm mark, side of the pipette was wiped with absorbent cotton and blood of the pipette was transferred immediately into specialized graduated tube for haemoglobin estimation containing 0.1N HCL At that time pipette was rinsed 2 or 3 times by sticking water and the washings were added to the solution in the same tube. The tube was shaked until the blood was well mixed with the hydrochloric acid and water and the mixture became uniformly dark brown in colour. After about 5 min. water was added into the brown colour solution drop by drop with a dropper and each time the solution was mixed with a stirrer until the color of the solution matches the standard color of haemoglobinometer. After matching the result was taken in day light from the scale of the measuring tube by observing the graduation mark at the lower edge of the meniscus at the top of the liquid column and expressed in g (%).

Erythrocyte Sedimentation Rate

Collected blood with anticoagulant was inserted into dry wintrobe haematocrit tube by pasteur pipette exactly up to 0 or 100 mark according to Barnhart (1969). Care was taken not to allow any bubble in the tube. The tube was placed in a special rack in vertical position for 1h, the erythrocyte sedimentation rate (ESR) was calculated by measuring the distance the erythrocyte had sedimented from the scale at the top of the tube and the result was expressed in mm/h.

RESULTS

During the period of investigation six (6) species of parasites of different groups could be collected and identified shown in a separate communication (Khalil *et al.* 2013). These parasites are *Euclinostomum multicaecum* Tubangui and Masilungun, 1935; *Allocreadium handiai* Pandey, 1937; *Lytocestus indicus* Moghe, 1925; *Pseudocaryophyllaeus heteropneustus* Chandra and Khatun, 1993; *Procamallanus heteropneustes* Ali, 1957 and *Paracamallanus equispiculus* Sood, 1968.

Effects of parasitism

Changes in the nature of growth (Length)

The experimental fishes were first differentiated as infested and uninfested and their average total length were presented in (Table 1) and their differences noted as 0.63 and the percentage loss of length was 3.39.

Table 1. The average length (TL) of uninfested and infested *H. fossils* and the percentage of loss of length

SL. No.	Infested or uninfested	Number Examined	Mean length (cm)	Loss of length (cm)	% Loss of length
1.	Uninfested	92	18.61±2.17	-	-
2.	Infested	118	17.98±2.21	0.63	3.39

Changes in the nature of growth (Weight)

During the period of investigation both the uninfested and infested host fish *H. fossilis* were weighted. The average weight of the uninfested hosts were 33.02±10.53 g and the average weight of the infested hosts were 29.77±11.88 g. Due to parasitic infestation the difference of weight was 3.25 g and the percentage loss of weight was 9.84 (Table 2). It appeared that there was a noticeable loss of weight of the host fish as a result of infestation of parasites after applying t-test at 5% level of significance.

Table 2. The average weight of uninfested and infested fish with the percentage loss of weight

SL. No.	Infested or Uninfested	Number Examined	Mean weight (g)	Loss of weight (g)	% Loss of weight
1	Uninfested	92	33.02±10.53	-	-
2	Infested	118	29.77±11.88	3.25	9.84

Changes in the nature of growth (Condition factor)

During the study a total of 210 *H. fossilis* were examined where 92 fish were uninfested and 118 fish were infested with parasites. The condition factor of uninfested and infested fishes presented in Table 3 and it was cleared that the uninfested fish have higher condition factor (0.53) than infested ones (0.51). From this results it was indicated there was a marked differences after applying t-test at 5% level of significance.

Table 3. Condition factor of uninfested and infested with parasites in *H. fossilis*

Infestation	Uninfested	Infested
Mean length (cm)	18.61±2.17	17.98±2.21
Mean weight (g)	33.02±10.53	29.77±11.88
Condition factor	0.53	0.51

Condition factor of uninfested and infested fish (*H. fossilis*) in different length Groups

The highest condition factors (0.65, 0.62 and 0.53) were found in uninfested fish in small, medium and large length groups than infested ones (0.62, 0.55 and 0.49) (Table 4). From this results it was indicated that there was a marked difference among different length groups after applying t-test at

5% level of significance. So, it was indicated that there were significant differences among different level of infestation among different level of infestation after applying t-test at 5% level of significance.

Changes in the nature of Gonad

A total 210 hosts were examined where 96 were male and 114 were female. The average weight of gonad in uninfested and infested male were 0.93±0.19 and 0.87±0.15. The percentage loss of gonad weight was 6.45. The average weight of gonad in uninfested and infested female were 1.57±0.95 and 1.50±0.71. The percentage loss of gonad weight in female was 4.46 (Table 5).

Table 4. Relationship between infestation and the condition factor of fish parasite in different length groups in *H. fossilis*

Length groups (cm)	12-16		17-21		22-25	
Infestation	Uninfested	Infested	Uninfested	Infested	Uninfested	Infested
Number of fish Examined	33	47	42	54	17	17
Mean length (cm)	14.52±1.21	14.41±1.01	17.58±1.28	18.25±1.18	21.52±1.55	21.44±0.56
Mean weight (g)	19.75±4.21	18.76±5.66	33.92±6.93	33.21±7.25	52.53±11.15	47.81±7.82
Condition factor	0.65	0.62	0.62	0.55	0.53	0.49
Loss of Condition factor (%)	-	4.62	-	11.29	-	7.55

Table 5. The average weight of gonad in infested and uninfested male and female

SL. No.:	Sex	No. of examined fish	Average Gonad weight of Uninfested fish	Average Gonad weight of infested fish	% Loss of gonad weight
01	Male	96	0.93±0.19	0.87±0.15	6.45
02	Female	114	1.57±0.95	1.50±0.71	4.46

Changes in the nature of blood composition

In this investigation it was observed that the hemoglobin content and Erythrocyte Sedimentation Rate (ESR) of uninfested and infested hosts. It was also observed in the percentage (%) of loss of hemoglobin and Erythrocyte Sedimentation Rate (ESR) at different level of infestation of different length groups.

Hemoglobin content in uninfested and infested *H. fossilis*

Total 210 hosts investigated to see the hemoglobin contents of *H. fossilis*. The range of hemoglobin percentage found in uninfested fishes was 2.8-6.2% and in infested it was found 2.2-6.2%. The average percentage in uninfested fishes was 4.77±1.02 and in infested was 4.18±0.71. The loss of hemoglobin content was 0.59% (Table 6). From the above results it was indicated that there was significant differences among infested and uinfested hosts after applying t-test at 5% level of significance.

Table 6. The hemoglobin content (%) of uninfested and infested fish with the percentage loss of hemoglobin

SL. No.	Infested or uninfested	Number Examined	Mean Hemoglobin (g/100ml)	% Loss of Hemoglobin
1	Uninfested	92	4.77±1.02	-
2	Infested	118	4.18±0.71	0.59

The average Hemoglobin (%), percentage loss of Hemoglobin in   different level of infestation of parasites in *H. fossils*

The highest (%) loss of hemoglobin were 0.77 found in larger length group than others group. The lowest % loss of hemoglobin were 0.35 found in medium length group. Overall differences in different infestation level of the host organism it was higher in high level of infestation in larger length group than others infection group (Table 7).

Table 7: The average Hemoglobin (%), percentage loss of Hemoglobin in   different level of infestation of parasites in *H. fossils*

Length groups (cm)	12-16			17-21			22-25		
Infestation	Uninfested	Low level of infestation (1-3) parasite	High level of infestation (more than 3) parasite	Uninfested	Low level of infestation (1-3) parasite	High level of infestation (more than 3) parasite	Uninfested	Low level of infestation (1-3) parasite	High level of infestation (more than 3) parasite
Number Examined	33	39	8	42	39	15	17	12	5
Average Hemoglobin (%)	4.67±1.02	4.04±0.66	3.97±0.68	4.68±1.13	4.33±0.70	4.20±0.94	4.98±0.84	4.47±0.34	4.21±0.23
% Loss of Hemoglobin	-	0.63	0.70	-	0.35	0.48	-	0.51	0.77

Erythrocyte Sedimentation Rate (ESR) in uninfested and infested *H. fossilis*

All fishes were examined to see the Erythrocyte Sedimentation Rate (ESR) of *H. fossilis*. The range of Erythrocyte Sedimentation Rate (ESR) percentage found in uninfested 1-6 mm/h and in infested it was found 1-6 mm/h. The average percentage in uninfested was 2.95±1.41 mm/h and in infested was 2.78±1.29 mm/h. The loss of ESR was 5.76% (Table 8).

Table 8. The ESR (%) of uninfested and infested fish with the percentage loss of ESR

SL. No.	Infested or uninfested	Number Examined	Mean ESR mm/hr	% Loss of ESR
01	Uninfested	92	2.95±1.41	-
02	Infested	118	2.78±1.29	5.76

ESR: Erythrocyte Sedimentation Rate

DISCUSSION

During the periods of investigation, six species of parasites *E. multicaecum, A. handiai, L. indicus, P. heteropneustus, P. heteropneustus, and P. equispiculus* were recorded from *H. fossilis*. These parasites were also reported by Ahmed *et al.* (1985), Sanaullah (1976), Mamnur Rashid *et al.* (1983) and Chandra and Khatun (1993) and Chandra (1994) from this host. So the present finding is the agreement of occurrence of these parasites in *H. fossilis* in Bangladesh waters specially from Sylhet region. However, Shahin *et al.* (2013) did not record caryophyllaeid parasites from *H. fossilis* in Mymensingh.

In the present investigation higher condition factor (0.53) recorded in uninfested fish than the average infested (0.51) ones. Das *et al.* (1997) described the mean values of condition factor and relative condition factor are 1.0755 and 1.0144 for catfish culture. When the values are in less than the mean values then the fishes fall in alarming situation due to parasitic infestation.

The highest loss of condition factor was found (11.29%) in medium length group fish. The lowest loss of condition factor was (4.62%) found in smaller length group. In *H. fossilis,* the condition factor decreases when the number of parasite increases. Similar finding was observed by Mann (1953) and Kabata (1958) in case of attack of *Lernaeocera*. Almost identical observations were made by Sproston and Hartely (1941).

In the present investigation the haemoglobin concentration ranged from 1.0 to 6.0 g/100 ml of blood and average haemoglobin concentration of uninfested fish was 4.77±1.02 g/100 ml and 4.18±0.71 g/100ml in infested hosts. The loss of haemoglobin percentage was 0.59. By using acid haematin method Schiffman and Formm (1959) found a haemoglobin value of 6.5 g/100 ml of blood for rainbow trout S. gairdneri. Mulcahy (1970) gave a range of 5.6 to 15.00 g/ml blood in pike determined through cyanmethaemoglobin method which is more or less similar to the present study.

Average erythrocyte sedimentation rate (ESR) of H. fossilis was 2.95±1.41 mm/h in uninfested fish and 2.78±1.29 mm/h in infested fish. Here the loss of ESR was 5.76% which is nearer to the result of Siddiqui and Naseem (1979) where the ESR value of L. ruhita was 2.5±0.98 mm/h. Blaxhall and Daisley (1973) and McCarthy et al. (1973) reported the ESR value of rainbow trout to be 1-5 mm/h and 1-8 mm/h respectively. In this study ESR value ranged from 1.0 to 6.0 mm/h in H. fossilis which is within the above ranges. However different plasma viscosity and specific gravity of erythrocyte are responsible for different ESR values.

REFERENCES

1. Ahmad G and G J Srivastava, 1985. Histopathological alternation in the liver and skin of *Heteropneustes fossilis* exposed chronologically to a sublethal concentration of methylene blue. Pakistan Journal of Zoology, 17 (3): 239-246.
2. Ali, 1957. *Procamallanus heteropneustes.* Indian Journal of Helminthology, 8, p. 3.
3. Barnhart R A, 1969. Effect of certain variables on haemetological characteristics of rainbow trout *Salmo gairdneri (R.).* Transactions of the American Fisheries Society, 98: 417-418.

4. Barua G, 1989. The status of epizootic ulcerative syndrome of fish of Bangladesh. In: R.J. Roberts, B. Campbell and I.H. Macrae (eds) ODA Regional Seminar on epizootic ulcerative syndrome. Aquatic Animal Health Research Institute, Bangkok, pp. 13-20.

5. Bhuiyan AL, 1964. Fisheries of Dacca, Published by Asiatic Society of Pakistan, Dacca, 1st ed. 148 pp.

6. Blaxhall PC and KW Daisley, 1973. Routine haematological methods for use with fish blood. Journal of Fish Biology, 5: 771–781.

7. Chandra KJ, 1993. Helminth parasites of certain freshwater and estuarine fishes of Bangladesh. Bangladesh Agricultural University Research Progress, 7: 643-654.

8. Chandra KJ and MR Khatun, 1993. A new species of caryophyllaeid cestode from *Heteropneustes fossilis* of Mymensingh. Rivista di Parassitologia, 10 (54): 235-239.

9. Chandra KJ, 1994. Infections, concurrent infections fecundity of *Procamallanus heteropneustes* Ali, parasitic to the fish *Heteropneustes fossilis*. Environment and Ecology, 12: 679-684.

10. Chandra KJ, NM Alam and MA Baki, 1996. Clinico-anatomical status on yellow grub disease of Singhi, *Heteropneustes fossilis*(Bloch) of Mymensingh. Bangladesh Journal of Agriculture, 21: 87-94

11. Chandra KJ and PC Modak, 1995. Activity, aging and penetration of the first stage larvae of *Procamallanus heteropneustes* Ali, 1957 (Nematode: Camallanidae). Asian Journal of Fisheries Science, 8: 95-101

12. Cross SX, 1993. Some host parasite relationship of the Trought lake region of Northern Wisconsis. Journal of Parasitology, 20 : 132-133.

13. Das NG, AA Majumder and S. M. M. Sarwar, 1997. Length-weight relationship and condition factor of catfish. Indian Journal of Fisheries, 44: 181-185.

14. Dias MLG, CV Mince-Vera, JC Eiras, MH Machado, GTT Souza and GC Pavnelli, 2006. Ecology of Clinostomum complanatum Rudolphi, 1814 (Trematoda : Clinostomidae) infecting fish from the floodplain of the high parna river, Brazil. Parasitological Research, 99(6): 675-681.

15. Hesser EF, 1960. Methods for routine fish haematology. Progessive Fish Culture, 522: 164-171.

16. Heinchiger RP, KD Laffarty, TC Huspeni, AJ Brookes and AM Kuris, 2007. Relationship between species richness and abundance of larval trematodes and of local benthos and fishes. Ecology Heildelburg Germany, 151(1): 82-92.

17. Hossain MA and M Shariff, 1992. Effect of copper on humoral immune response in gold fish, *Carassius auratus* L. In: M. Shariff. R. Subasinghe and J. R. Arthur (eds.). *Disease in Asian Aquaculture, Manila,* Fish Health Section, Asian Fisheries Society, pp. 371-377.

18. Kabata Z, 1958. *Lernaeocera obtusa* n. sp. it biology and its effects on the haddock. Marine Research, 3: 1-26.

19. Kabata Z, 1985. *Parasites and diseases of fish cultured in the tropics.* Taylor and Francis Ltd. pp. 318.

20. Khalil SMI, KJ Chandra, MT Hasan, MA Kashem and DR Das, 2013. Parasitic infestations and their influence on length and weight of the host, *Heteropneustes fossilis*. Bangladesh Journal of Environmental Science, 24: 114-119.

21. Laboni NN, KJ Chandra and MS Chhanda, 2012. Effects of caryophyllaeid cestode infestations on *Clarias batrachus* (Linn.). Journal of Asiatic Society of Bangladesh, Science, 38 (2): 135-144.

22. Mackiewicz JS, 1982. Caryophyllidean (Cestoidea): Perspectives. Parasitology, 84: 397-417.

23. Moghe MA, 1925. A supplementary description of *Lytocestus indicus* Moghe, Cestoda. Parasit, 23: 84-87.

24. Mulcahy MF, 1970. Blood values in the pike *Esox lucius* L. Journal of Fish Biology, 2: 203–209.

25. Mamnur Rashid M, AKM Aminul Haque and KJ Chandra, 1983. Records of some metazoan parasites of *Clarias batrachus* (Linnaeus) from Mymensingh. Bangladesh Journal of Fisheries, 6: 37-42.

26. Mann H, 1953. *Lernaeocera branchialis* (*Copepoda Parasitica*) and seine schadwirkung bei einigen Gadiden. Archa Fischereiwiss, 1952/53: 133-143.

27. McCarthy DH, JP Stevenson and MS Roberts, 1973. Some blood parameters of the rainbow trout *(Saltno gairdneri,* Richardson.). Journal of Fish Biology, 5: 1-8.

28. Pandey KC, 1937. Studies on Cestodes of India. III on a new species of *Allocreadium handiai*, 932 from *Heteropneustes* (Ham.). Indian Journal of Zoology, 4: 54-55.

29. Sanaullah M, 1976. Contribution to the studies of some metazoan parasites in *Heteropneustes fossilis* (Bloch) and *Clarias batrachus* (Linnaeus) in Bangladesh. M. Sc. Thesis, Dept. of Zoology, University of Dhaka. pp. 116.

30. Schiffman RH and PO Fromm, 1959. Measurement of some physiological parameters in rainbow trout, *Salmo gairdneri (P.)*. Cananadian Journal of Zoology, 37: 25-32.

31. Shahin MIH, KJ Chandra, DR Das and SMI Khalil 2013. Morphology and histopathology of alimentary canal of *Clarias batrachus* (Linnaeus) and *Heteropneustes fossilis* (Bloch). International Journal of Applied Life Sciences, 2(3): 11-20.

32. Smith GC, WM Lewis and HM Kaplan, 1952. A comparative morphologic and physiologic study of fish blood. Progressive Fish Culture, 14: 169-172

33. Sood, 1968. *Paracamallanus equispiculus*. Indian Journal of Helminthology, 20(2). pp. 91.

34. Siddiqui AQ and Naseem SM, 1979. The haematology of Rohu, *Labeo rohita*. Journal of Fisheries Biology, 14: 67–72.

35. Sproston NG and PHT Hartely, 1941. The ecology of some parasitic copepods of gadoids and othe fishes. Journal of Marine Biological Association, U.K., 25: 361-417.

36. Tubangui MA and F Masilungun, 1935. Caryophyllaeid parasites from *H. fossilis* of Philippine VI. Descriptions of new species and classification. Philippines Journal of Sciences, 51:167-174.

22

LENGTH-WEIGHT RELATIONSHIP AND GROWTH PATTERN INFERENCE OF A SMALL INDIGENOUS FRESHWATER PRAWN, *Macrobrachium lamarrei* (H. Milne-Edwards, 1837) IN BANGLADESH

Motia Gulshan Ara, Md. Noor Nabi, Zoarder Faruque Ahmed and Mst. Kaniz Fatema[*]

Department of Fisheries Management, Faculty of Fisheries, Bangladesh Agricultural University, Mymensingh-2202, Bangladesh

*Corresponding author: Mst. Kaniz Fatema, E-mail: kanizhossain@gmail.com

---

ARTICLE INFO	ABSTRACT

Key words:

Freshwater prawn
Length
Weight
Growth pattern

The length-weight relationships (LWRs) of *Macrobrachium lamarrei* (H. Milne-Edwards, 1837) was studied to construct the generalized relationships of body weight to carapace length measurements for male, female and combined prawn population separately over a calendar year. A total of 1018 specimens were used for this study. The parameter values of the equation $W = a L^b$ describing the relationships between body weight and carapace length for male, female and combined prawns. The parameter of equations, a varied from 0.00096 to 0.00333, 0.00059 to 0.00728 and 0.00081 to 0.00440 for male, female and combined prawn population respectively. On the other hand, the slope of equations, b varied from 2.4363 to 2.8229, 2.0761 to 3.0314 and 2.2915 to 2.8998 for male, female and combined prawn population respectively indicating isometric growth in females but negative allometric for males suggesting that female prawns were comparatively in better condition than males of the same population. The generalized length-weight relationship was fitted with the pooled data of all monthly samples for male, female and combined prawn separately which were $BW = 0.0008\ CL^{2.924}$, $BW = 0.0012\ CL^{2.783}$ and $BW = 0.0010\ CL^{2.845}$ respectively indicating negative allometric growth as b varied from 2.783 to 2.924. The results further revealed that all length-weight relationships (LWRs) were highly correlated ($r^2 > 0.904$).

INTRODUCTION

The prawns of Palaemonidae family are very important economic resources in the world's crustacean fishery (Mantelatto and Barbosa, 2005; Hossain *et al.*, 2012). Palaemonid prawn has also immense importance in Bangladesh as it considered one of the most important sectors of the national economy. Among different prawn species available in Bangladesh, M. *lamarrei* is notable one, commonly known as "kuncho river prawn" that is a small indigenous freshwater prawn belonging to the family Palaemonidae. This kuncho river prawn has found in ponds and rivers in Bangladesh, India and some other countries (Holthius, 1980). Previously, a few studies were conducted on *M. lamarrei* based on their general biology. However, in Bangladesh there is hardly any work that has been conducted on the growth and length-weight relationships (LWR) in this very valuable Palaemonid prawn species.

Length-weight relationship is very important for establishing production and biomass estimation of a species. (Anderson and Gutreuter, 1983). Sarkar *et al.* (2008) and Mir *et al.* (2012) reported that the estimation of the average weight of the fish of a given length group by establishing a mathematical relation between length-weight. The relationships between length and weight are important in fisheries management for comparative growth studies (Hossain *et al.*, 2006). Besides this in setting yield equations for comparing the population in space and time and estimating the number of fish landed, length-weight relationship can also be used (Singh *et al.*, 2011). Length and weight data are a useful tool for standard result of fish sampling programs. Length- weight relationships are useful for estimating standing crop biomass and seasonal variations in fish growth can also be tracked by L-W relationship (Martin-Smith, 1996). Therefore, for assessing fish stock, length- weight relationship of fish species is important and for length-weight conversion, the parameters 'a' and 'b' can be used (Pauly and Gayanilo, 1996). Furthermore, the empirical relationship between the length and weight of the fish enhances the knowledge of natural history of commercially important fish species, thus making the conservation possible.

Subsequently, the aims of the present paper is to determine the length-weight relationships of monthly samples, to construct a generalized length-weight relationship of *M. lamarrei* for male, female and combined prawn separately over a calendar year and to evaluate the growth pattern of male, female and combined populations from L shaped pond at Bangladesh Agricultural University campus, Mymensingh, Bangladesh. Therefore, the present study provides baseline information for the growth of this species. The data and information will go a long way in supporting the management initiatives for the sustainability of this freshwater prawn as well as other species.

MATERIALS AND METHODS

Sample collection

M. lamarrei prawn species was collected randomly using a fine-meshed push net from the L-shaped pond situated in Bangladesh Agricultural University campus, Mymensingh, Bangladesh once in a month over a calendar year. All specimens were preserved in 10% buffered formalin soon after collection.

Sex determination

Male was identified observing second chelate legs which were extraordinary large, spiny and strong with naked eyes. Individuals confirmed as female with the presence of one pair of genital pore in the coax of third pairs of walking legs under a stereo microscope (Figure 1 and 2).

Recording of carapace length and body weight

Carapace length (CL) was measured from the tip of the rostrum to the posterior edge of the carapace with a slide calipers to the nearest 0.01 mm, whilst body weight (BW) was taken with a digital balance to the nearest 0.01 g. Detailed description of prawn sampling is given in Table 1.

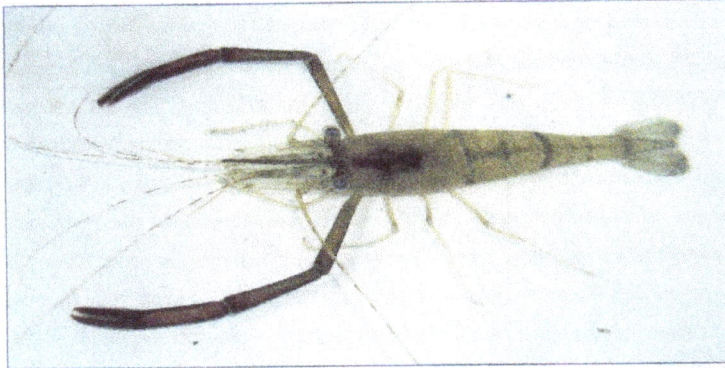

Figure 1. An image of a male freshwater prawn *M. lamarrei* (H. Milne-Edwards, 1837)

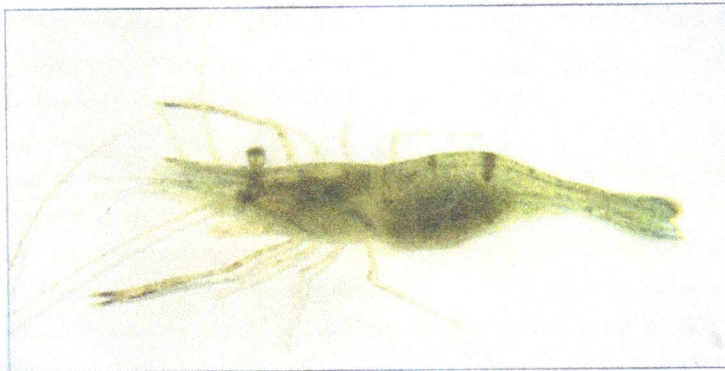

Figure 2. An image of a female freshwater prawn *M. lamarrei* (H. Milne-Edwards, 1837)

Length - weight relationship (LWR)

The relationship between carapace length and weight was calculated using the equation, $W = a\,CL^b$ where coefficient 'a' is the intercept in the y-axis, and the regression coefficient 'b' is an exponent indicating isometric growth when equal to 3. This equation was converted into its logarithmic expression: $\ln W = \ln a + b \ln CL$ where the parameters 'a' and 'b' were calculated by linear regression, as was the coefficient of determination (r^2). To test for possible significant differences in both slope and intercept, we followed the analysis of covariance (ANCOVA). All statistical analyses were considered significant at $P < 0.05$.

Growth pattern inference

Growth pattern of individuals in monthly sample were determined on the basis of b value in the equation $BW = a\,CL^b$. When 'b' is equal to 3, the growth was isometric resulting an ideal shape of prawn; when the value of *b* was less than 3.0, the growth was negative allometric and when the value of *b* was more than 3.0, the growth was positive allometric pattern. The growth pattern of the individuals in the population was evaluated judging the confidence interval of 'b' value at 95% confidence level.

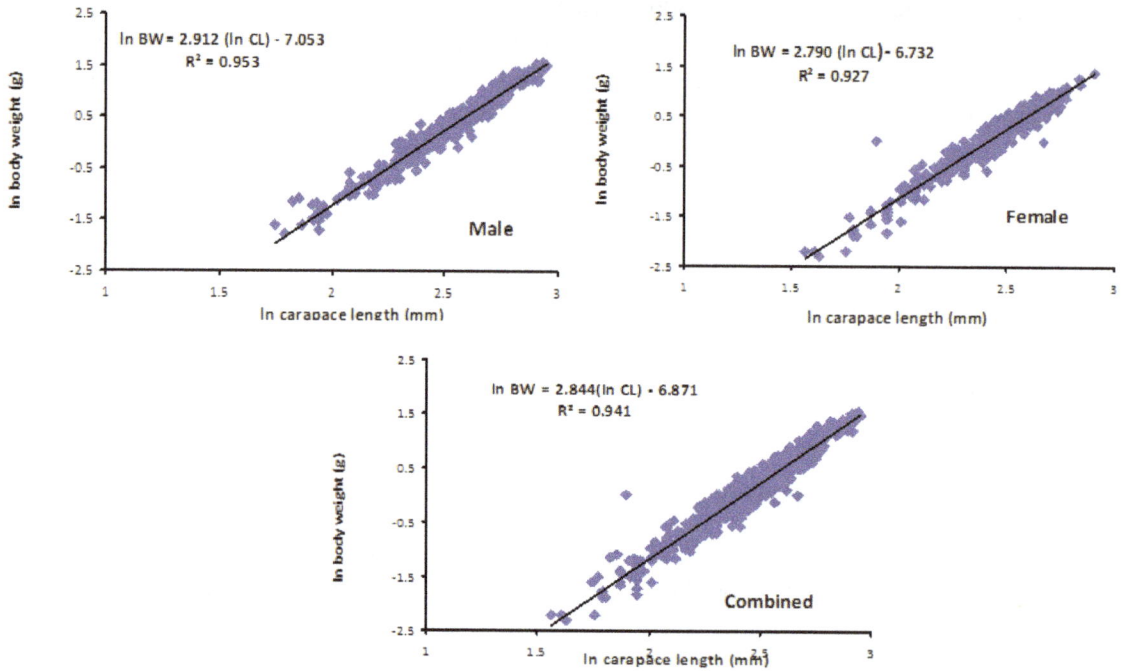

Figure 3. The generalized relationships between the natural logarithm of body weight (g) and the natural logarithm of carapace length (mm) in male, female and combined prawn of *Macrobrachium lamarrei*.

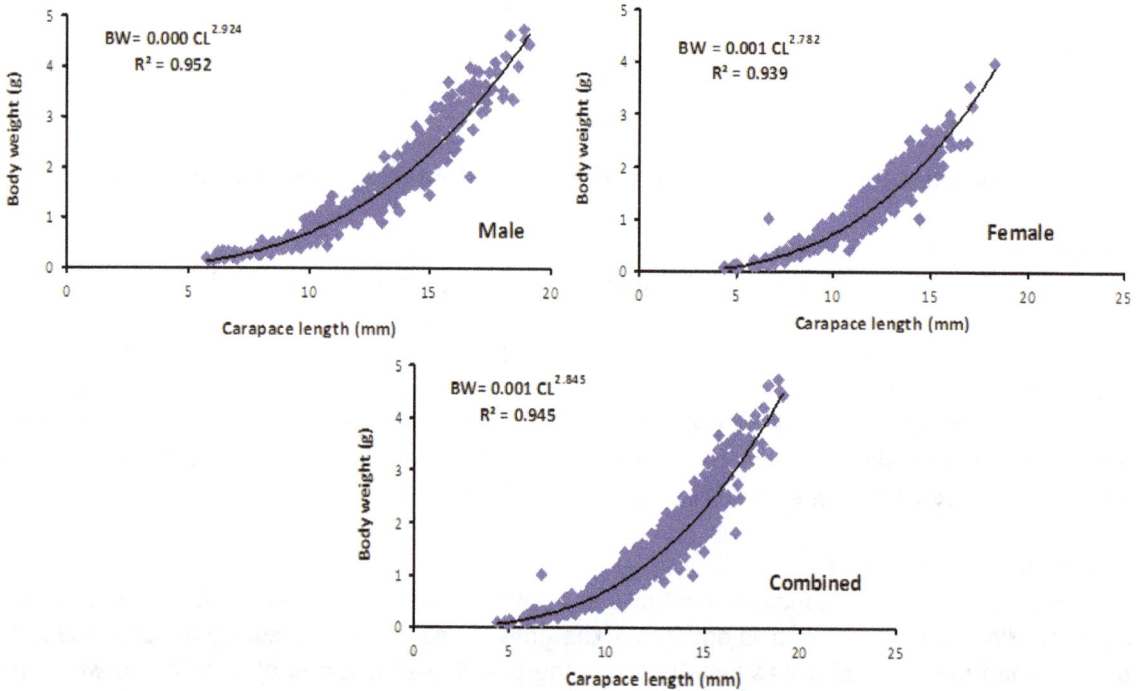

Figure 4. The generalized relationships between the body weight (g) and the carapace length (mm) in male, female and combined prawn of *Macrobrachium lamarrei*.

RESULTS AND DISCUSSIONS

The carapace length and body weight of all monthly samples collected all over the year varied and ranged from 4.80 mm to 19.10 mm and from 0.10 g to 4.76 g respectively. A total of 1018 specimens of *M. lamarrei* were used to determine the length- weight relationship where 485 were male and 533 were female prawns showing a sex ratio 1: 1.09 (male: female). The carapace length and body weight of male ranged from 5.75 mm to 19.10 mm and from 0.17 g to 4.76 g respectively. And the carapace length and body weight of female ranged from 4.80 mm to 18.29 mm and the body weight ranged from 0.10 g to 3.98 g (Table 1).

Table 1. Collection record of freshwater prawn *M. lamarrei* from the L shaped pond at Bangladesh Agricultural University (BAU), Mymensingh

Sampling Date	Number of Male	Size Range		Number of Female	Size Range	
		CL(mm)	BW(g)		CL(mm)	BW(g)
January	39	5.75-16.00	0.18-2.36	48	4.80-15.60	0.11-2.36
February	26	10.00-18.00	0.76-3.60	41	6.68-15.14	0.45-2.50
March	37	6.22-18.48	0.32-4.08	24	11.05-18.29	1.10-3.98
April	42	9.80-19.10	0.81-4.76	32	9.70-15.15	0.86-2.51
May	43	10.93-17.31	1.07-3.65	34	5.89-16.05	0.22-2.52
June	26	6.78-18.92	0.30-4.64	39	7.63-15.45	0.41-2.73
July	35	8.24-15.66	0.41-2.82	53	9.22-15.35	0.46-2.65
August	43	7.12-15.35	0.25-2.11	71	7.00-15.50	0.21-2.44
September	44	6.00-15.30	0.17-2.11	53	8.30-15.80	0.31-2.83
October	55	6.00-13.10	0.17-1.38	51	5.00-13.80	0.10-1.93
November	41	6.40-18.00	0.34-3.53	34	7.60-17.10	0.42-3.17
December	54	7.00-15.00	0.26-2.20	53	8.00-17.00	0.46-3.56

By establishing the relationship of body weight to carapace length different 'a' and 'b' values were found in this study for male, female and combined prawn separately. In case of male, monthly variation was found in coefficients of equations 'a' and it was from 0.00096 to 0.00333. The maximum and minimum values of 'a' were on March and September. The slope of equations 'b' of male also varied monthly and it was from 2.4363 to 2.8229. The maximum and minimum values of 'b' were on September and March. In case of female, monthly variation was found in coefficients of equations 'a' and it was from 0.00059 to 0.00728. The maximum and minimum values of 'a' were on January and June. The slope of equations 'b' of female also varied monthly and it was from 2.0761 to 3.0314. The maximum and minimum values of 'b' were on June and January. This result revealed that the growth in females were isometric but negative allometric in males suggesting that female prawns were comparatively in better condition than males of the same population.

Table 2. Monthly descriptive statistics and estimated parameters of length–weight relationships of *M. lamarrei* prawn species from a large perennial pond at Bangladesh Agricultural University, Mymensingh

Month	Sex	n	a	b	$S_e(b)$	95% CI of a	95% CI of b	r^2	Growth Inference
January	Male	39	0.00151	2.67713	0.09078	0.0011-0.0023	2.4932-2.8611	0.95919	Allometric
	Female	48	0.00104	2.80382	0.07238	0.0008-0.0015	2.6581-2.9495	0.97026	Allometric
	Combined	87	0.00110	2.78924	0.05485	0.0009-0.0014	2.6802-2.8983	0.96818	Allometric
February	Male	26	0.00156	2.69836	0.13116	0.0008-0.0031	2.4277-2.9690	0.94634	Allometric
	Female	41	0.00728	2.07607	0.15752	0.0034-0.0156	1.7575-2.3947	0.81664	Allometric
	Combined	67	0.00440	2.29152	0.10361	0.0026-0.0073	2.0846-2.4984	0.88271	Allometric
March	Male	37	0.00327	2.43627	0.10228	0.0019-0.0057	2.2286-2.6439	0.94189	Allometric
	Female	24	0.00267	2.53837	0.18698	0.0011-0.0072	2.1506-2.9261	0.89336	Allometric
	Combined	61	0.00363	2.40504	0.08430	0.0023-0.0056	2.2364-2.5737	0.93242	Allometric
April	Male	42	0.00185	2.66422	0.10148	0.0011-0.0032	2.4591-2.8693	0.94515	Allometric
	Female	32	0.00342	2.43162	0.18002	0.0013-0.0088	2.0641-2.7993	0.85879	Allometric
	Combined	64	0.00216	2.60819	0.07813	0.0014-0.0033	2.4524-2.7639	0.93931	Allometric
May	Male	43	0.00129	2.79799	0.10942	0.0007-0.0023	2.5770-3.0191	0.94100	Allometric
	Female	34	0.00149	2.72730	0.09532	0.0009-0.0024	2.5331-2.9215	0.96238	Allometric
	Combined	77	0.00129	2.79262	0.05634	0.0011-0.0017	2.6804-2.9048	0.97038	Allometric
June	Male	26	0.00184	2.67344	0.14587	0.0008-0.0041	2.3724-2.9745	0.93331	Allometric
	Female	39	0.00134	2.78644	0.15746	0.0006-0.0029	2.4674-3.1055	0.89434	Allometric
	Combined	65	0.00142	2.76582	0.09022	0.0009-0.0023	2.5855-2.9461	0.93717	Allometric
July	Male	35	0.00105	2.80381	0.11113	0.0006-0.0018	2.5777-3.0299	0.95071	Allometric
	Female	53	0.00059	3.03145	0.15581	0.0003-0.0013	2.7187-3.3442	0.88127	Isometric
	Combined	88	0.00088	2.87554	0.08944	0.0006-0.0014	2.6977-3.0533	0.92319	Allometric

Table 2. Contd.

Month	Sex	n	a	b	$S_e(b)$	95% CI of a	95% CI of b	r^2	Growth Inference
August	Male	43	0.00114	2.73272	0.09232	0.0007-0.0018	2.5463-2.9192	0.95531	Allometric
	Female	71	0.00077	2.93153	0.09812	0.0005-0.0012	2.7358-3.1273	0.92825	Allometric
	Combined	114	0.00081	2.89983	0.07301	0.0006-0.0011	2.7552-3.0445	0.93372	Allometric
September	Male	44	0.00096	2.82288	0.09215	0.0006-0.0015	2.6369-3.0088	0.95716	Allometric
	Female	53	0.00067	2.98841	0.09676	0.0004-0.0011	2.7942-3.1827	0.94925	Isometric
	Combined	97	0.00082	2.89806	0.06739	0.0006-0.0011	2.7643-3.0318	0.95115	Allometric
October	Male	55	0.00159	2.62406	0.06238	0.0012-0.0021	2.4989-2.7492	0.97092	Allometric
	Female	51	0.00128	2.73920	0.07639	0.0009-0.0018	2.5857-2.8927	0.96329	Allometric
	Combined	106	0.00145	2.67235	0.04945	0.0012-0.0018	2.5743-2.7704	0.96562	Allometric
November	Male	41	0.00221	2.53743	0.06925	0.0016-0.0031	2.3974-2.6775	0.97177	Allometric
	Female	34	0.00227	2.53001	0.11811	0.0013-0.0040	2.2894-2.7706	0.93481	Allometric
	Combined	75	0.00225	2.53063	0.06102	0.0017-0.0030	2.4090-2.6522	0.95929	Allometric
December	Male	54	0.00122	2.79011	0.10109	0.0007-0.0020	2.5872-2.9931	0.93611	Allometric
	Female	53	0.00174	2.65731	0.04818	0.0014-0.0022	2.5606-2.7540	0.98351	Allometric
	Combined	107	0.00152	2.70523	0.05102	0.0012-0.0021	2.6041-2.8064	0.96399	Allometric

n: number of specimens, a: coefficient parameter of length-weight relationship, b: slope of length-weight relationship , $S_e(b)$: standard error of b, CI: confidence interval, r^2: coefficient of determination

In case of combined prawn, monthly variation was found in coefficients of equations 'a' and it was from 0.00081 to 0.00440. The maximum and minimum values of 'a' were on February and August. The slope of equations 'b' of combined prawn also varied monthly and it was from 2.2915 to 2.8998. The maximum and minimum values of 'b' were on August and February. This result agrees with the finding of Soomro et al. (2012) for *Macrobrachium malcolmsonii.* They found the values of b were 2.97 for males and 3.04 for females indicating the growth was isometric for female and allometric for male.

Table 2 illustrated that the length-weight relationships of male *M. lamarrei* prawn species indicated negative allometric growth in all months over a calendar year. Similarly in female prawns, the length-weight relationships indicated negative allometric growth in all months except July and September where the growth was isometric. But in combined population the length-weight relationships indicated negative allometric growth for all the months over a calendar year. Finally, this study constructed the generalized length-weight relationship of all monthly samples for male, female and combined prawn that were $BW = 0.0008 \, CL^{2.924}$, $BW = 0.0012 \, CL^{2.783}$ and $BW = 0.0010 CL^{2.845}$ respectively. Lalrinsanga et al. (2012) reported similar results for *M. Rosenbergii* prawn and the length weight relationship were $W = 0.073299 \, L^{3.5502}$, $W = 0.102962 \, L^{3.2443}$ and $W = 0.087694 \, L^{3.3893}$ for male, female and total prawn respectively. In the present study, analysis of covariance revealed significant differences between sexes for the slopes 'b' of the regression lines ($P < 0.05$), with all coefficient of determination values being >0.904.

CONCLUSION

Length-weight relationship (LWR) is an important fishery management tool (Bagenal and Tesch, 1978; Gonzalez Acosta et al., 2004). They are used for various aspects including estimation of weight from length observations, calculation of production and biomass in the assessment of fish populations, in addition providing information about body condition of specimens in stocks or populations (Lai and Helser, 2004; Gerritsen and McGrath, 2007). Information regarding any biological aspects of *M. lamarrei* prawn species from Bangladesh waters is quite insufficient. To the best of our knowledge, no references dealing with length-weight relationship (LWR) for *M. lamarrei* is available from Bangladesh waters. Therefore, this study presents the first attempt to describe this length-weight relationship for the *M. lamarrei* species from Bangladesh. In conclusion, this study has provided the first basic and baseline information on the LWR that would be beneficial for fishery biologists and conservationists to impose adequate regulations for sustainable fishery management and conservation of biodiversity for this species in Bangladesh. Further, the information provides a platform for future research into other growth and reproductive attributes of this fish species.

REFERENCES

1. Anderson R and S Gutreuter, 1983. Length, weight and associated structural indices. In: *Fisheries techniques* (eds. L. Nielson and D. Johnson.). American Fisheries Society, Bethesda. pp. 283–300.
2. Bagenal TB and FW Tesch, 1978. Age and growth. In: Methods for Assessment of Fish Production in Fresh Waters, 3rd edit. (Begenal T, ed.). Blackwell Science Publications, pp.101-136.
3. Gerritsen HD and D McGrath, 2007. Significance differences in the length-weight relationships of neighbouring stocks can result in biased biomass estimates: Examples of haddock (*Melanogrammus aeglefinus*, L.) and whiting (*Merlangius merlangus*, L.). Fisheries Research, 85: 106-111.
4. Gonzalez Acosta AF, GD Aguero and JD Lacruz Aguero, 2004. Length-weight relationships of fish species caught in a mangrove swamp in the Gulf of California (Mexico). Journal of Applied Ichthyology, 20: 154-155.

5. Holthius, LB 1980. FAO Species Catalogue, shrimps and prawns of the world. An annoted catalogue of species of interest to fisheries. FAO Fisheries Synopsis, 125 271p.

6. Hossain MY, ZF Ahmed, PM Leunda, AKMR oksanul Islam, S Jasmine, ZJ Osco, R Miranda and J Ohtomi, 2006. Length-weight and length-length relationships of some small indigenous fish species from the Mathabhanga River, southwestern Bangladesh. Journal of Applied Ichthyology, 22: 301-303.

7. Hossain MY, J Ohtomi, A Jaman, J Saleha and LVJ Robert, 2012. Life history traits of the Monsoon River prawn *Macrobrachium malcolmsonii* (Milne-Edwards, 1844) (Palaemonidae) in the Ganges (Padma) River, northwestern Bangladesh. Journal of Freshwater Ecology, 27: 131-142.

8. Lai HL and T Helser, 2004. Linear mixed-effects models for weight-length relationships. Fisheries Research, 70: 377-387.

9. Lalrinsanga PL, RP Bindu, G Patra, S Mohanty, NK Naik and S Sahu, 2012. Length weight relationship and condition factor of giant freshwater prawn *Macrobrachium rosenbergii* (De Man, 1879) based on developmental stages, culture stages and sex. Turkish Journal of Fisheries and Aquatic Sciences, 12: 917-924.

10. Mantelatto FLM and LR Barbosa, 2005. Population structure and relative growth of freshwater prawn *Macrobrachium brasiliense* (Decapoda, Palaemonidae) from São Paulo State, Brazil. Acta Limnologica Brasiliensis, 17: 245-255.

11. Martin-Smith KH, 1996. Length- weight relationship of fishes in a diverse tropical freshwater community, Sabah, Malaysia. Journal of Fish Biology, 49: 731-734.

12. Mir JI, UK Sarkar, AK Dwivedi, OP Gusain, A Pal and JK Jena, 2012. Pattern of intrabasin in condition factor, relative condition factor and form factor of an Indian Major Carp, *Labeo rohita* (Hamilton- Buchanan, 1822) in the Ganges Basin, India. European Journal of Soil Biology, 4: 126-135.

13. Pauly D and FC Jr. Gayanilo, 1996. Estimating the parameter of length-weight relationship from length-frequency samples and bulk weights. 23 321p.

14. Sarkar UK, PK Deepak and RS Negi, 2008. Length- weight relationship of clown knife fish *chitala chitala* (Hamilton, 1822) from the Ganga basin, India. Journal of Applied Ichthyology, 25: 232-233

15. Singh NO, D Sarma and NG Singh, 2011. Length- weight relationship of *Tor putitora* (Hamilton) from Kosi River, Uttarakhand considering different stages of its lifespan. Indian Journal of Fisheries, 58: 35-38.

16. Soomro AN, WA Baloch, TJ Chandio, WM Achakzai and S Saddozai, 2012. Condition factor and length-weight relationship of monsoon river prawn *Macrobrachium malcolmsonii malcolmsonii* (H. Milne-Edwards, 1844) (Palaemonidae) in lower Indus river. Pakistan Journal of Zoology, 44: 1279-1283.

Permissions

All chapters in this book were first published in RALF, by AgroAid Foundation; hereby published with permission under the Creative Commons Attribution License or equivalent. Every chapter published in this book has been scrutinized by our experts. Their significance has been extensively debated. The topics covered herein carry significant findings which will fuel the growth of the discipline. They may even be implemented as practical applications or may be referred to as a beginning point for another development.

The contributors of this book come from diverse backgrounds, making this book a truly international effort. This book will bring forth new frontiers with its revolutionizing research information and detailed analysis of the nascent developments around the world.

We would like to thank all the contributing authors for lending their expertise to make the book truly unique. They have played a crucial role in the development of this book. Without their invaluable contributions this book wouldn't have been possible. They have made vital efforts to compile up to date information on the varied aspects of this subject to make this book a valuable addition to the collection of many professionals and students.

This book was conceptualized with the vision of imparting up-to-date information and advanced data in this field. To ensure the same, a matchless editorial board was set up. Every individual on the board went through rigorous rounds of assessment to prove their worth. After which they invested a large part of their time researching and compiling the most relevant data for our readers.

The editorial board has been involved in producing this book since its inception. They have spent rigorous hours researching and exploring the diverse topics which have resulted in the successful publishing of this book. They have passed on their knowledge of decades through this book. To expedite this challenging task, the publisher supported the team at every step. A small team of assistant editors was also appointed to further simplify the editing procedure and attain best results for the readers.

Apart from the editorial board, the designing team has also invested a significant amount of their time in understanding the subject and creating the most relevant covers. They scrutinized every image to scout for the most suitable representation of the subject and create an appropriate cover for the book.

The publishing team has been an ardent support to the editorial, designing and production team. Their endless efforts to recruit the best for this project, has resulted in the accomplishment of this book. They are a veteran in the field of academics and their pool of knowledge is as vast as their experience in printing. Their expertise and guidance has proved useful at every step. Their uncompromising quality standards have made this book an exceptional effort. Their encouragement from time to time has been an inspiration for everyone.

The publisher and the editorial board hope that this book will prove to be a valuable piece of knowledge for researchers, students, practitioners and scholars across the globe.

List of Contributors

Rabeya Yesmin, Salina Akhter Sume, Md. Nazmul Haque and Golam Quader Khan
Department of Fisheries Biology and Genetics, Bangladesh Agricultural University, Mymensingh-2202, Bangladesh

Nargis Sultana
Department of Fisheries Biology and Aquatic Environment, Bangabandhu Sheikh Mujibur Rahman Agricultural University, Salna, Gazipur-1706, Bangladesh

Gias Uddin Ahmed, Md. Towhid Hasan, Md. Ali Reza Faruk and Md. Nazmul Hoque
Department of Aquaculture, Faculty of Fisheries, Bangladesh Agricultural University, Mymensingh-2202, Bangladesh

Md. Khalilur Rahman
Bangladesh Fisheries Research Institute, Mymensingh- 2202, Bangladesh

Gias Uddin Ahmed, Habiba Aktar, Sudristi Chakma, Neaz Al Hasan and Mohammad Shamsuddin
Department of Aquaculture, Faculty of Fisheries, Bangladesh Agricultural University, Mymensingh-2202, Bangladesh

Salma Noor-E-Islami
Department of Fisheries (DoF), Government of the People's of Republic Bangladesh

Md. Faisal
Department of Fishing and Post Harvest Technology, Faculty of Fisheries, Chittagong Veterinary and Animal Sciences University, Chittagong-4225

Mousumi Akter
Department of Fisheries Technology, Faculty of Fisheries, SFM Fisheries College, Melandah, Jamalpur-2010

Md. Shaheed Reza and Md. Kamal
Department of Fisheries Technology, Faculty of Fisheries, Bangladesh Agricultural University, Mymensingh-2202, Bangladesh

Matouke Matouke Moise
Department of Biological Sciences, Federal University, Dutsinma, Katsina, Nigeria

Obadiah Audu Abui
Department of Biological Sciences, Federal Science and Technical College, Kafanchan, Kaduna State, Nigeria

Subrata Kumar Ghosh, Mirja Kaizer Ahmmed and Sk. Istiaque Ahmed
Faculty of Fisheries, Chittagong Veterinary and Animal Sciences University, Chittagong-4225, Bangladesh

Md. Khabirul Ahsan and Md. Kamal
Faculty of Fisheries, Bangladesh Agricultural University, Mymensingh-2202, Bangladesh

Md. Anisur Rahman, Flura, Tayfa Ahmed, Md. Mehedi Hasan Pramanik and Mohammad Ashraful Alam
Bangladesh Fisheries Research Institute, Riverine Station, Chandpur-3602, Bangladesh

Md. Shahjahan
Department of Fisheries Management, Faculty of Fisheries, Bangladesh Agricultural University, Mymensingh-2202, Bangladesh

Sk. Istiaque Ahmed
Department of Fisheries Management, Faculty of Fisheries, Bangladesh Agricultural University, Mymensingh-2202, Bangladesh
Faculty of Fisheries, Chittagong Veterinary and Animal Sciences University, Khulshi, Chittagong-4225, Bangladesh

Mirja Kaizer Ahmmed, Subrata Kumar Ghosh and Md. Moudud Islam
Faculty of Fisheries, Chittagong Veterinary and Animal Sciences University, Khulshi, Chittagong-4225, Bangladesh

Ibrahim Shehu Jega
Department Fisheries Management, Faculty of Fisheries, Bangladesh Agricultural University, Mymensingh-2202, Bangladesh

Ibrahim Mohammed Ribah
Department of Animal Science, Kebbi State University of Science and Technology, Aliero, Nigeria

Zakariyya Idris
Department of Forestry and Fisheries, Kebbi State University of Science and Technology, Aliero, Nigeria

M. Mahfujul Haque, Md. Saifullah Bin Aziz and Md. Mehedi Alam
Department of Aquaculture, Faculty of Fisheries, Bangladesh Agricultural University, Mymensingh-2202, Bangladesh

Md. Mahbubur Rahman and Md. Nurunnabi Mondal
Department of Fisheries Management, Bangabandhu Sheikh Mujibur Rahman Agricultural University, Gazipur, Bangladesh

Jannatun Shahin
Department of Fisheries, Ministry of Fisheries and Livestock, Bangladesh

Jannatul Fatema
Department of Agricultural Economics

Mst. Kaniz Fatema
Department of Fisheries Management, Bangladesh Agricultural University, Mymensingh-2202, Bangladesh

Tasnuba Hasin
Department of Fisheries Resources Management, Faculty of Fisheries, Chittagong Veterinary and Animal Sciences University, Chittagong-4225, Bangladesh

Baadruzzoha Sarker
Production Officer, BRAC Fish Hatchery, Sreemangal, Moulvibazar-3214, Bangladesh

Muhammad Forhad Ali
Department of Aquaculture, Sheikh Fajilatunnesa Mujib Fisheries College, Melandah, Jamalpur, Bangladesh

Md. Masud Rana and Subhash Chandra Chakraborty
Department of Fisheries Technology, Faculty of Fisheries, Bangladesh Agricultural University, Mymensingh-2202, Bangladesh

Apurba Roy
Department of Economics, Faculty of Social Sciences, University of Barisal, Barisal-8200, Bangladesh

Md. Abdus Samad, Mst. Lutfunnahar and Md. Selim Reza
Department of Fisheries, University of Rajshahi, Rajshahi-6205, Bangladesh

Sujit Kumar Chatterjee
Fish inspection and quality control officer, DOF, Dhaka, Bangladesh

Md. Ashrafuzzaman
SUFO, DOF, Panchagarh Sadar, Bangladesh

Mst. Mansura Khan
Department of Fisheries Biology and Genetics, Faculty of Fisheries, Bangladesh Agricultural University, Mymensingh-2202, Bangladesh

Mohammad Sadequr Rahman Khan
Department of Marine Bioresources Science, Chittagong Veterinary and Animal Sciences University, Chittagong-4225, Bangladesh

Shahrear Hemal, Md. Shahab Uddin and Md. Tariqul Alam
Department of Aquaculture, Sylhet Agricultural University, Sylhet-3100, Bangladesh

Md. Saif Uddin
Department of Aquaculture, Bangabandhu Sheikh Mujibur Rahman Agricultural University, Gazipur-1706, Bangladesh

Bhaskar Chandra Majumdar and Md. Golam Rasul
Department of Fisheries Technology, Bangabandhu Sheikh Mujibur Rahman Agricultural University, Gazipur-1706, Bangladesh

Sharmin Akter Shampa, Nusrat Nasrin, Marufa Khatun and Salma Akter
Department of Fisheries Biology and Genetics, Faculty of Fisheries, Bangladesh Agricultural University, Mymensingh-2202, Bangladesh

Sonia Sku, Md. Golam Sajed Riar and Nur-A-Raushon
Freshwater Station, Bangladesh Fisheries Research Institute, Mymensingh-2201, Bangladesh

Sumit Kumer Paul and Md. Kamal
Department of Fisheries Technology, Faculty of Fisheries, Bangladesh Agricultural University, Mymensingh-2202, Bangladesh

Md. Amzad Hossain and Mousumi Das
Department of Aquaculture, Bangabandhu Sheikh Mujibur Rahman Agricultural University (BSMRAU), Gazipur-1706, Bangladesh

Md. Shahanoor Alam
Department of Genetics and Fish Breeding, BSMRAU, Gazipur-1706, Bangladesh

Md. Enamul Haque
Department of Agriculture Extension and Rural development, BSMRAU, Gazipur-1706, Bangladesh

Sarker M. Ibrahim Khalil
Department of Fish Health Management, Sylhet Agricultural University, Sylhet-3100 Bangladesh

Kirtunia Juran Chandra
Department of Aquaculture, Bangladesh Agricultural University, Mymensingh-2202, Bangladesh

David Rintu Das
Bangladesh Fisheries Research Institute, Freshwater Sub Station, Saidpur, Nilphamari, Bangladesh

Motia Gulshan Ara, Md. Noor Nabi, Zoarder Faruque Ahmed and Mst. Kaniz Fatema
Department of Fisheries Management, Faculty of Fisheries, Bangladesh Agricultural University, Mymensingh-2202, Bangladesh

Index